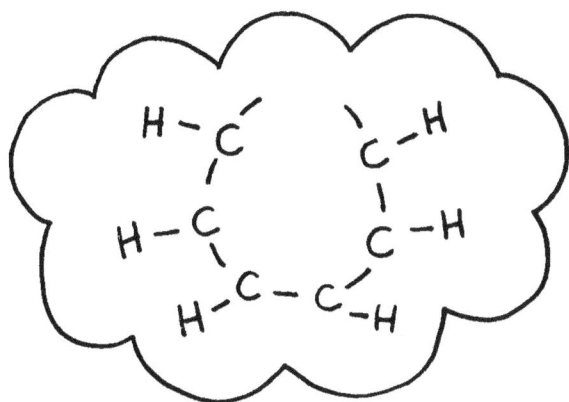
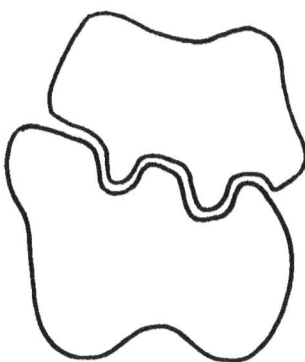
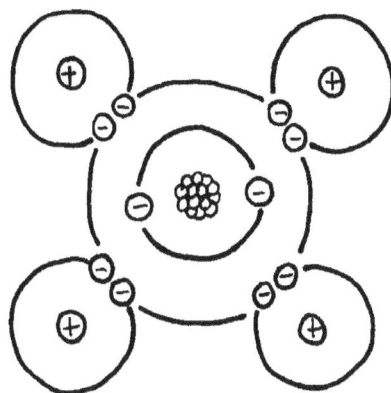

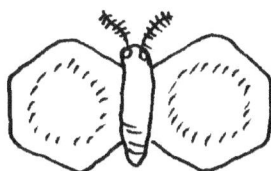

# CARBON

# CHEMISTRY

## Second Edition — Student text

## An introduction to organic chemistry

by Ellen J. McHenry

© 2021  Ellen Johnston McHenry

ISBN 978-1-7374763-2-0

(Student text for Carbon Chemistry)

The author gives permission to the purchaser
to make copies as needed for use
with a classroom or homeschool group.

For queries about other uses:
ejm.basementworkshop@gmail.com

# Table of Contents

# CHAPTER 1: CARBON ATOMS

The heart of carbon chemistry is, of course, the carbon atom.  Like all atoms, the carbon atom is made of only three particles: protons, neutrons, and electrons.  There are several ways to represent a carbon atom.  Each model has strengths and weaknesses.

This is called the **electron cloud model**. It shows the areas where the electrons "live" around the nucleus (center) of the atom. It shows us roughly what an atom actually looks like. However, it is almost useless when we want to study the orderly arrangement of electrons into shells and orbitals, or when we want to show chemical bonding.

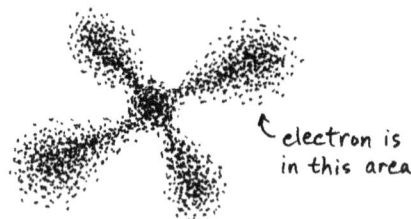

This is called the **solar system model**.  It doesn't look anything like a real carbon atom, but it is a very good model to use for learning about the arrangement of protons, neutrons and electrons.  It helps us to understand how the electrons orbit around the nucleus.  We can show the arrangement of the electrons into shells and easily count them. The weakness of this model is that it doesn't look at all like a real atom.

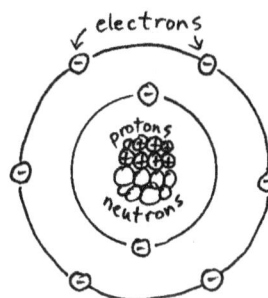

This is called the **ball and stick model**.  It doesn't look like a real carbon atom, either.  The ball in the center represents both the nucleus of the atom, and any electrons that are in "inner" shells, closer to the nucleus. The sticks represent free electrons on the outside of the atom that are available for bonding with other atoms.  This model is very useful when you want to build models of molecules.  It does not show the electrons, however; it shows only sticks where the bonds are, and this can be confusing to beginning students.  You have to remember that the stick represents an electron or a pairing up of electrons.

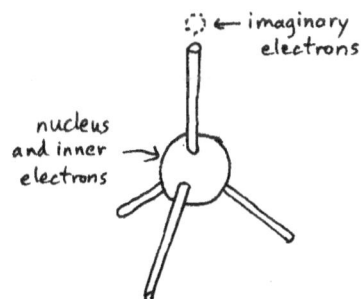

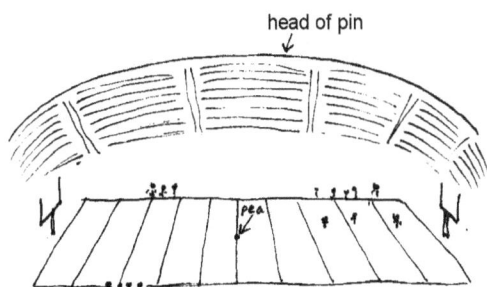

A weakness of all these models is that they do not show the relative sizes and distances between the particles.   If you imagine that the nucleus of an atom is a marble sitting on the 50 yard line inside a large football stadium, the electrons would be pin heads traveling along the outer reaches of the upper decks.  It's hard to believe, but an atom is mostly empty space!

1

Each element has a unique number of protons. Hydrogen has one proton, helium has two, lithium has three, beryllium has four, and so on through the Periodic Table. An atom's atomic number tells how many protons it has. Carbon's atomic number is six, so it has six protons.

*Not too much for me to complain about yet. Doesn't seem too hard so far.*

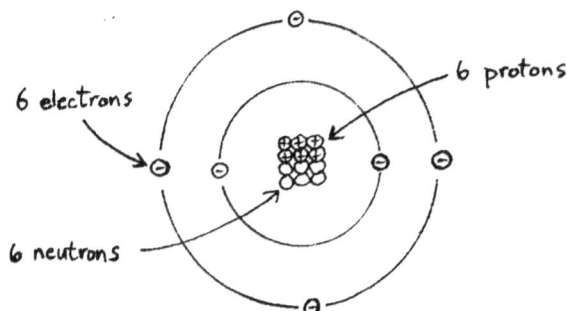

6 electrons

6 protons

6 neutrons

The plus signs in the protons mean that they carry a positive electrical charge. The minus signs in the electrons show that they have a negative charge.

Since atoms must be electrically balanced, this also means that carbon has six electrons. Carbon's electrons are arranged in two layers, or shells. The first shell contains two electrons, and the remaining four are in the second shell. The fact that carbon has four electrons in its outer shell is very significant. Ideally, all atoms would like to have their outer shells filled, and, in the case of carbon, it would like to have eight electrons, not four. Like most of the smaller atoms on the Periodic Table, carbon lives by the motto: **"8 is great!"**

Electrons form pairs, with one electron spinning one way, and the other electron spinning in the opposite direction. (Imagine a very simple dance.) Carbon would like each of its electrons to have a partner, so carbon is out looking for four electron "dance partners" to fill in these empty places.

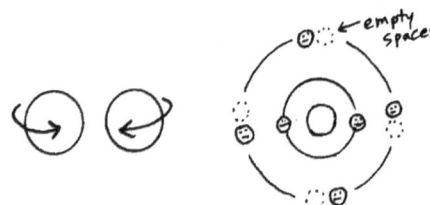

empty spaces

How does carbon find electrons to fill in these empty places? It borrows them from other atoms. It just so happens that there are other atoms out there that have the same problem that carbon does. They have electrons without partners, too. These atoms would love to get together with carbon and share one or more electrons, in an attempt to make pairs of electrons. Let's look at some atoms that would like to share electrons with carbon.

Hydrogen is the smallest atom that exists. It is made of only one proton and one electron. What fun can just one electron have? The proton isn't much company—it can't do the electron dance. So, hydrogen's electron goes out looking for a partner.

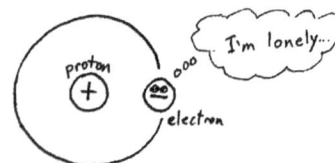

proton

I'm lonely...

electron

Look! There's a carbon atom. It needs some partners! So hydrogen goes over to carbon and puts its electron into one of carbon's empty slots. Now we have one happy electron couple.

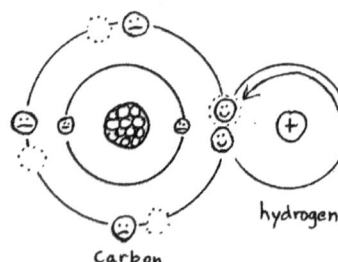

hydrogen

Carbon

Then the hydrogen atom calls up three of its hydrogen friends and tells them to come on over and fill the other three slots. Now we have a real square dance. Carbon is thrilled to have partners for its four electrons. This works out rather well!

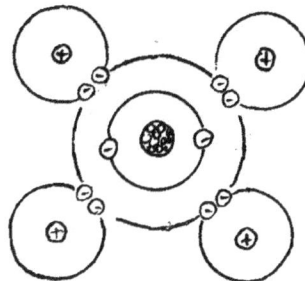

Another atom that can cooperate with carbon is chlorine. Chlorine's problem is that it has seven, not eight, electrons in its outer shell. Chlorine is out looking for a free electron that can pair up with its lonely electron. Can carbon do this? Carbon has four electrons that are looking for partners. One of those electrons could go over and fill in chlorine's empty slot. What if there were four chlorines that were all looking for partners and they were all willing to come over to the carbon and pair up with one of carbon's lonely electrons? Hey—this works out pretty well, too!

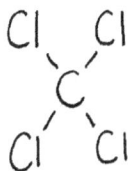

Here is an easy way to draw it.

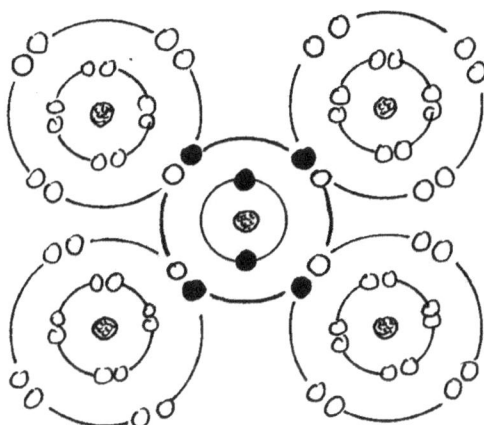

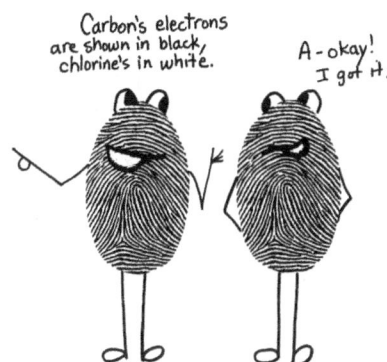

Carbon's electrons are shown in black, chlorine's in white.

A-okay! I got it!

Life is seldom perfect, even in the atomic kingdom. Sometimes things don't work out so well and carbon must adapt to unusual "dance partners." For example, sometimes carbon has to make do with only two atoms, not four. In the carbon dioxide molecule, carbon pairs up with two oxygens. Since oxygen has two free partners, two oxygens can provide a total of four partners—just what carbon is looking for. All they need to do is slide their electrons over a bit and make them match up. (Electrons don't really "slide" over, but what they do is too complicated for this discussion.)

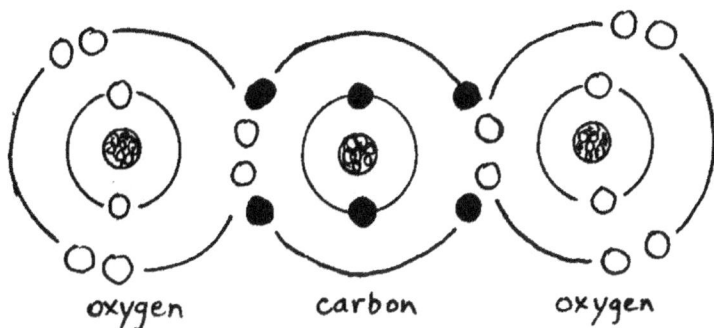

oxygen          carbon          oxygen

Oxygen pretends two of carbon's electrons belong to it, so it also has eight electrons in its outer shell.

We can draw it like this:

$$O=C=O$$

When carbon doubles up like this, we call it a **double bond**. That makes sense, doesn't it?

Carbon can also bond with itself. The only problem is that the carbon atoms on the edges will have unpaired electrons hanging off. Nevertheless, carbon does bond with itself. The free electrons dangling on the edges usually pick up a hydrogen atom, or some other atom that happens to be in the area.

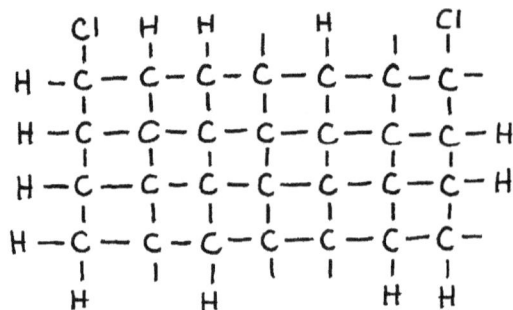

There are basically three ways that carbon bonds with itself. Each of these substances is called an **allotrope**. The first allotrope of carbon is **diamond**. Diamonds are made of pure carbon. The bonds between the carbons are extremely strong, making diamond the hardest substance on earth. Diamonds are so hard they can be used on industrial saw blades to cut metal and concrete. This picture shows how the carbon atoms are linked in diamonds.

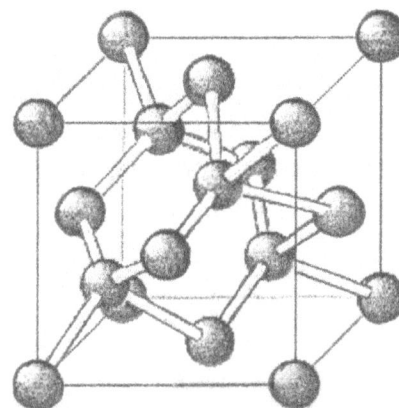

Historically, the only way to get diamonds was dig them up out of the earth. The largest diamond mines in the world are located in southern Africa, Russia, and western Australia. (Some of these mines have treated their workers very poorly, which has caused diamond mining to become controversial.) Fortunately, diamonds can now be manufactured artificially, so industries that need diamonds do not need to rely on controversial mining companies. Artificial diamonds often start out as methane, a substance we'll meet very soon. Small, thin diamond plates are put into a high-pressure cooker along with the methane gas. The heat and pressure causes the carbon atoms to stick to the carbon plate and a diamond crystal begins to grow. If you'd like to see some videos about this process, go to www.youtube.com/thebasementworkshop, click on "playlists," then find the "Carbon Chemistry" playlist. (You might have to click on "show all playlists" if you don't see it listed.)

Another allotrope of carbon is **graphite**. You use graphite all the time; it is the "lead" in pencils. Real lead (Pb on the Periodic Table) is not used in pencils anymore, because it was discovered to be dangerous to our health. You can see that in graphite, the carbon atoms are arranged in layers. Each layer is made of a sheet of hexagonal shapes. The layers are loosely bonded to each other and can slide around. This is what makes graphite feel slippery. If you rub your fingers on the end of a pencil, the slippery sensation you feel is the graphite layers sliding back and forth. Because it is slippery, graphite can be used as a dry lubricant. Some people rub a pencil on the drawer runners in dressers so that the drawers go in and out smoothly.

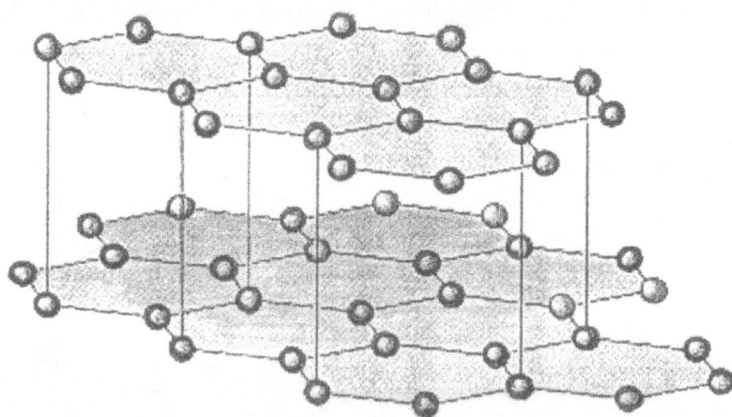

It's hard to believe that graphite and diamonds are made of the same stuff, but if you could squeeze your pencil tip hard enough (which you can't), the atoms would rearrange themselves to form a diamond!

A single sheet of graphite is called **graphene**. Graphene has recently been discovered to have amazing properties. It has been called the strongest substance in the world, even stronger than diamonds or steel. It is the best conductor of electricity and it's also transparent and flexible. Many industries are developing uses for graphene. If it sounds too good to be true, that may indeed turn out to be the case, however. It might turn out be one of the most destructive substances on earth if it gets released into the environment. We don't know enough about it yet.

The third allotrope was not discovered until 1985. It was named **buckminsterfullerene**, after the architect Buckminster Fuller, who was famous for his geodesic dome structures in the 1960s and '70s. Since the name is so long, scientists have come up with a nickname for this substance. They call the molecules **buckyballs**.

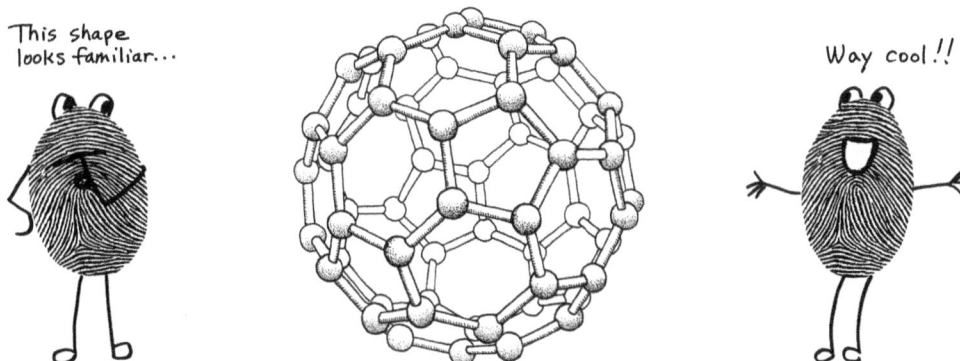

This shape looks familiar...

Way cool !!

If you think this pattern looks like a soccer ball, you're right—the pattern is the same. There are 20 hexagons and 12 pentagons, with each pentagon completely surrounded by hexagons.

What are buckyballs good for? Some scientists think they might be good for microscopic lubrication, or bearings in a microscopic motor. They might be used inside the human body for drug delivery (by putting molecules of medicine inside the buckyballs). If you add a few potassium atoms to the buckyball, it will conduct electricity as well as metal does. At low temperatures, it becomes a superconductor.

Where can you find these weird balls? Buckyballs are a component of black soot—the kind that collects on the glass screen in front of a fireplace. Scientists don't go around collecting soot, however. They manufacture buckyballls in their labs by vaporizing graphite with a laser.

Two more forms of carbon that should be mentioned are **coal** and **charcoal**. They are made of mostly carbon, but often have impurities such as nitrogen, sulfur, salt, or rock and dust particles. (When the sulfur comes out into the air as coal is burned, it can cause major air pollution problems.) In coal, the carbon atoms are not bonded into geometrical shapes. The scientific word for "no shape" is **amorphous**. ("A" means "without," and "morph" means "shape.") Coal and charcoal are said to be amorphous types of carbon. Coal seems to have come from ancient plants that were buried and then put under extreme pressure. Charcoal is made by burning wood in a low-oxygen environment.

FORMATION OF COAL    (under intense pressure)

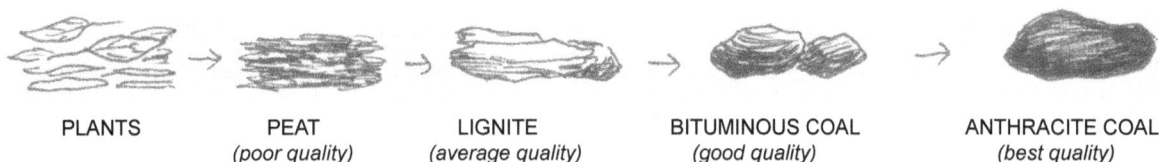

| PLANTS | PEAT | LIGNITE | BITUMINOUS COAL | ANTHRACITE COAL |
|---|---|---|---|---|
| | (poor quality) | (average quality) | (good quality) | (best quality) |

5

## Comprehension self-check

See if you can fill in the blanks with the correct words.  If you have trouble remembering, go back and re-read that section and find the answer.

1)  All atoms are made of three types of particles:  _____, _____, and _____.

2)  The three types of atomic models mentioned in this chapter are _____, _____, and _____.

3)  Which model gives us the best picture of what an atom really looks like? _____

4)  Which model is the best one to use when making molecule models? _____

5)  Which model is the best for showing exactly what is going on with the arrangement of electrons into shells? _____

6)  Which one is easiest to draw? _____

7)  Which one is easiest to build out of craft materials? _____

8)  If we were to make a model of an atom that was proportionately correct, our nucleus would be the size of a _____ in a _____ and the electrons would be the size of _____ traveling around the _____.

9)  It is the number of _____ that make an atom what it is.  This number is called the _____ number.

10)  Most atoms in the top part of the Periodic Table (the smaller, non-metal atoms) live by this motto: "_____"

11)  If an atom does not have a full outer shell of electrons, what does it do about it? _____ _____

_____

12)  When carbon has to double up and share more than one electron with another atom, we call this a _____ bond.

13)  Three substances that demonstrate how carbon atoms bond with each other in geometrical shapes are _____, _____, and _____.

14) Name a use for graphite other than in pencils: _____.

15) A carbon "buckyball" is patterned in the same way as a _____.

16)  A single layer of graphite is called _____.

17) What does "amorphous" mean? _____

18) Name two amorphous forms of carbon: _____ and _____

19) Coal is made of _____ that lived a long time ago.

20) Charcoal is made by burning _____ in a low-oxygen environment.


## A sample from "The Chemical Elements Coloring and Activity Book"

The following two pages are from of this author's advanced coloring book about the elements of the Periodic Table.  If you'd like to have all 118 elements, you can find the book by searching ISBN 978-1-7374763-0-6  at your favorite online book store (Amazon, Barnes & Noble, BooksAMillion, BookDepository, etc.).

# C

# Carbon

*From the Latin word for charcoal: "carbo"*

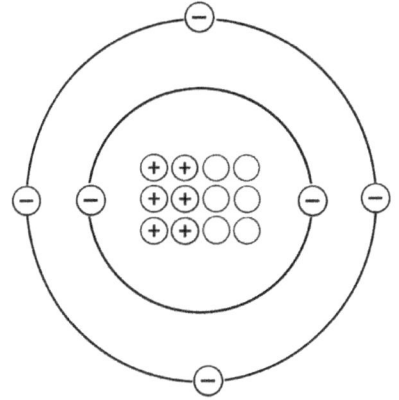

6 protons
6 neutrons
6 electrons

Atomic mass: 12.01

Carbon is the most flexible and "friendly" atom on the Periodic Table. It will bond with many other elements, although its favorites are hydrogen and oxygen. If there are no other atoms around to bond with, carbon will bond to itself, forming pure-carbon substances such as diamonds, graphite and coal. That's right, coal and diamonds are made of the same stuff! The most fascinating pure-carbon structure is the buckyball, a hollow sphere of 60 carbon atoms arranged in the same pattern as a soccer ball (hexagons surrounded by pentagons).

Carbon is found in the air around us as carbon dioxide, $CO_2$. Vehicles can put both $CO_2$ and CO (carbon monoxide) into the air as by-products of combustion. CO is very dangerous and many people have CO detectors in their homes if they have a furnace that burns natural gas, $CH_4$.

Carbon can bond to three oxygen atoms and make the carbonate ion, $CO_3^{2-}$. If a calcium atom sticks to carbonate, we get calcium carbonate, $CaCO_3$. Calcium carbonate is the main ingredient in the mineral calcite and in the rock known as limestone. Sea shells are a biological form of calcium carbonate.

Hydrocarbon molecules are made of just carbon and hydrogen atoms and can be small ($CH_4$, natural gas), medium-sized ($C_8H_{18}$, octane, liquid gasoline) or so long we can't even count the carbon atoms (plastics and rubbers). Carbon and hydrogen atoms can also form a ring known as benzene. The benzene ring, or an adaptation of it, is at the heart of thousands of molecules, including polystyrene plastic, Styrofoam®, food preservatives, cholesterol, natural almond flavor, spot removers, moth balls, paints, and medicines.

Many biological molecules have carbon at their core. Proteins, fats and sugars are all carbon-based substances. DNA, the extremely long ladder-shaped molecule that is like a library of information for living cells, has carbon atoms at key points in its structure. Carbon is also at center of many other molecules essential to life, including enzymes.

**Octane $C_8H_{18}$**

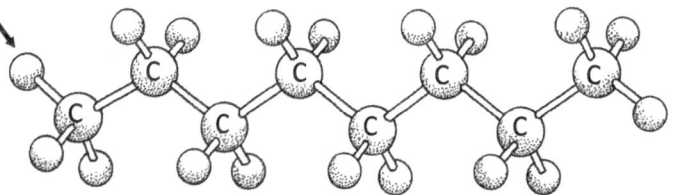

**Limestone $CaCO_3$**

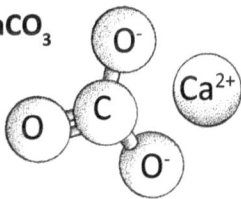

**Methane $CH_4$**

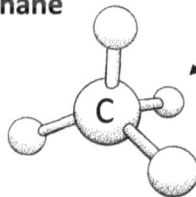

These are all H

**Benzene $C_6H_6$**

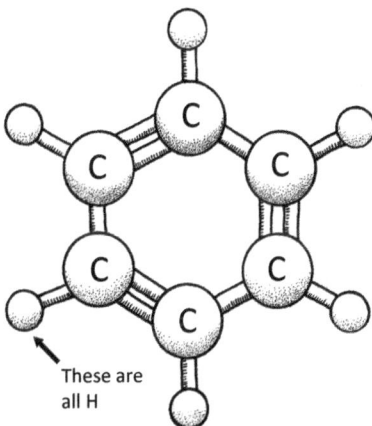

These are all H

**Diamond lattice (pure C)**

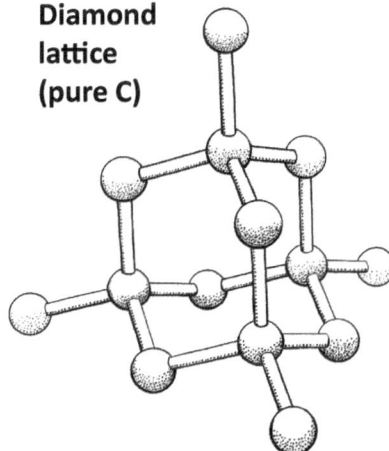

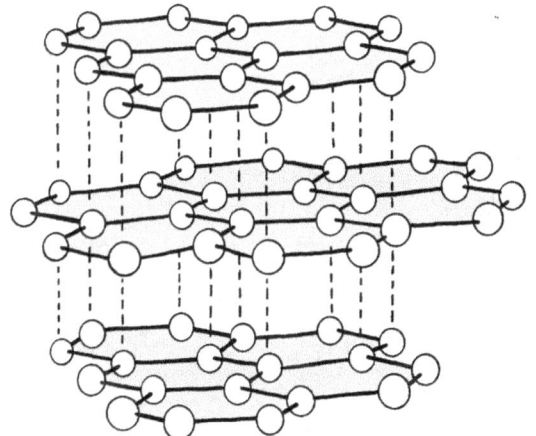

**Graphite lattice (pure C)**

# 6 Carbon C

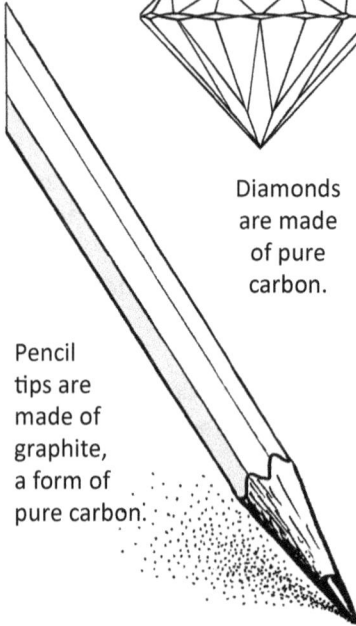

Diamonds are made of pure carbon.

Pencil tips are made of graphite, a form of pure carbon.

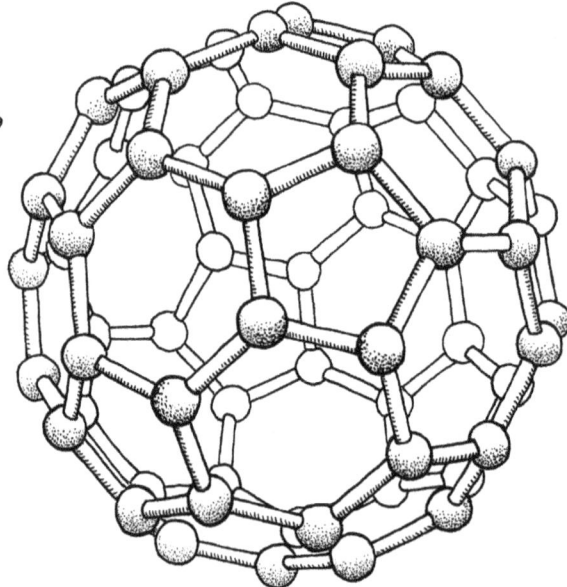

"Buckyballs" are made of 60 carbon atoms.

Carbon atoms determine the structure of DNA.

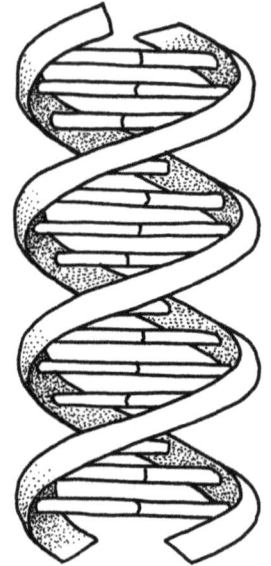

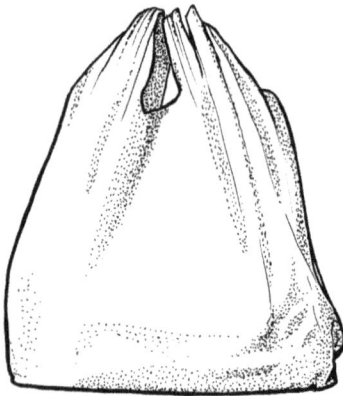

All forms of plastic are made of long chains of carbon atoms (with hydrogens attached).

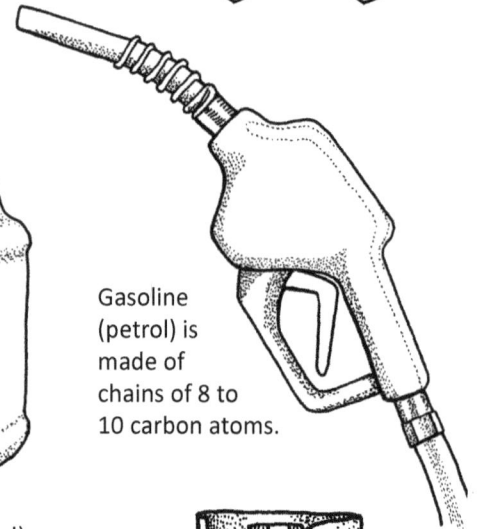

Gasoline (petrol) is made of chains of 8 to 10 carbon atoms.

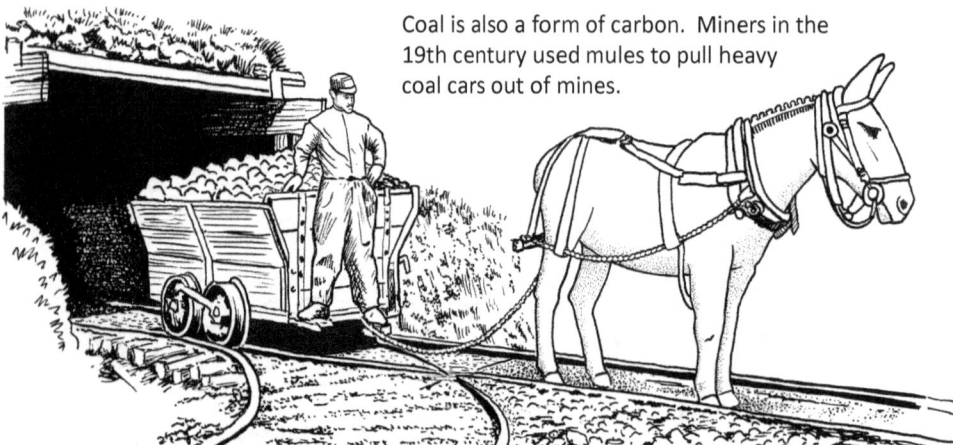

Coal is also a form of carbon. Miners in the 19th century used mules to pull heavy coal cars out of mines.

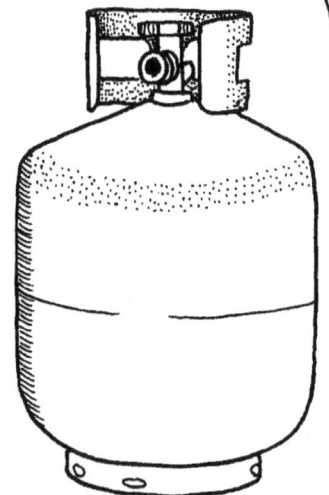

Natural gas (methane) is $CH_4$.

# CHAPTER 2: ALKANE HYDROCARBONS

We learned in chapter one that carbon often bonds with hydrogen. When carbon bonds with just hydrogen, they form a molecule we call a **hydrocarbon**. The simplest hydrocarbon is **methane**. It consists of one carbon atom and four hydrogen atoms:

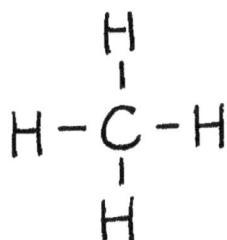

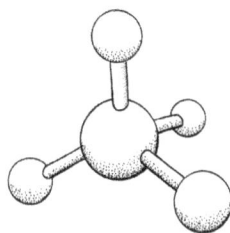

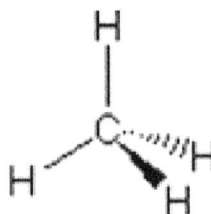

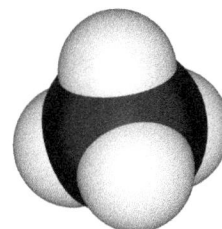

Here are four ways of drawing the methane molecule. The first one (on the far left) is called a "structural formula." It is easy to draw, and because it uses letters to represent the atoms, you always know what the atoms are. However, it does not show the three-dimensional shape of the molecule. The second one is the "ball and stick" model. The hydrogen atoms want to stay as far apart from each other as possible, and this "tetrahedral" shape is the result. The third model is a hybrid between the structural model and the ball and stick. The last model is a "space-filling" model. It probably comes closest to showing us what a real methane molecule looks like. Space-filling models are easy to make out of clay, but are difficult to draw. We won't be using them very much in this book, but it is good to have seen one and know what it is called.

Methane is a small, lightweight molecule that floats around in the air as a gas. You can't see it or smell it. We sometimes call it "natural gas" because it occurs naturally in the earth, often forming in areas where oil and coal are found. Natural gas can be used to heat homes, but the gas companies must add a smelly substance to it so that we will be aware of any leaks in the pipes.

Methane burns easily in the presence of oxygen, and it burns cleanly, without polluting the air. This makes it excellent for use as a fuel, but it also makes coal mining a dangerous job. Miners can run into pockets of methane gas as they work. A spark of any kind could ignite the gas and create a deadly mine fire. In the early days of mining, the miners sometimes took caged birds with them into the mines. The birds were very sensitive to the methane gas and would act strangely, or even faint, if there was methane present. By watching the behavior of the bird, the miners would have an early warning signal telling them that methane was lurking in the mine.

Where was the S. P. C. A.?!

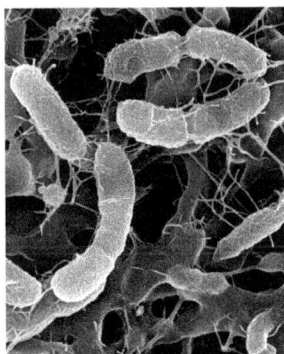

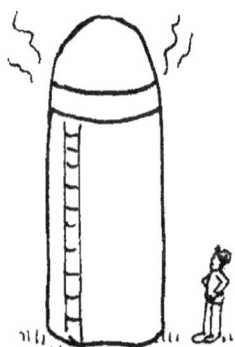

Some types of bacteria produce methane. A good place to find these bacteria is in rotting vegetation. Silos used by farmers to store chopped corn and hay will also contain a healthy population of bacteria that eat the plants and produce methane as a waste product. Care must be taken to monitor how much methane is building up inside the silo. High levels of methane can lead to spontaneous silo fires.

Methane-producing bacteria live in our intestines, also. Yes, gas is really… gas. Healthy intestines have millions of harmless (and beneficial) bacteria living in them. We need these bacteria in our intestines. They aid in digestion and keep us healthy. When certain foods pass through the intestines undigested, the bacteria produce an extra amount of methane and hydrogen. But remember, methane has no odor. The odor we associate with intestinal gas comes from very small amounts of other substances such as hydrogen sulfide. Since methane is flammable, it is fortunate that the methane in our intestines is mixed with other gases such as nitrogen and carbon dioxide, which are not flammable. However, there is enough methane in some intestines to cause problems. Surgeons in the early days of medicine learned the hard way about the flammability of methane when sparks from their operating instruments would occasionally cause small explosions in the patients' intestines!

Let's leave that last paragraph unillustrated and move on!

Methane is the first and simplest member of a whole group of carbon compounds called **alkanes**. The second member of this group has two carbon atoms in it and is called **ethane**. This is how it looks:

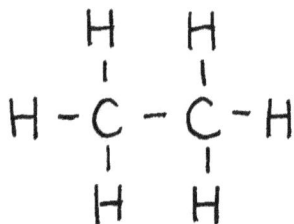

structural formula          the ball and stick model          space filling model

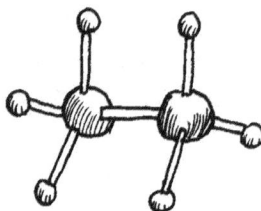

As you can see, ethane is made of two carbon atoms and six hydrogen atoms. We can write it like this: $C_2H_6$. (This is called the **empirical formula**.) Both carbon atoms have all four of their free electrons attached to another atom, so this combination works out well. Like methane, ethane is a gas.

If another carbon atom is added on, we make a substance called **propane** ($C_3H_8$).

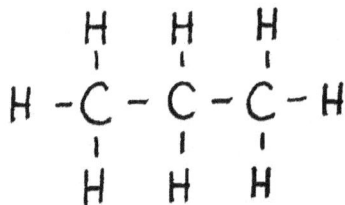

Can we stop drawing those other two models? Thanks!

Undoubtedly, you've heard of propane. You may have a propane tank outside your house, connected to a gas grill.

Add another carbon atom to the string, plus a few more hydrogens, and you have a molecule named **butane** ($C_4H_{10}$). Butane can be found in hand-held lighters.

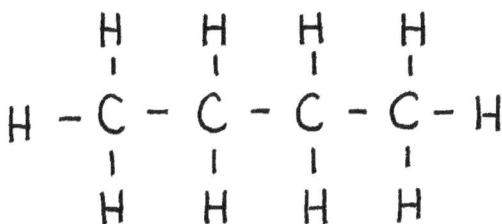

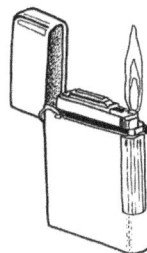

Butane is a heavy gas and easily liquified. Butane in lighters and torches will usually be a liquid.

You can keep on adding carbon atoms and make the string longer and longer. You could have dozens, hundreds, thousands, or millions of carbon atoms in an alkane string. Short strings with 1 to 4 carbon atoms are gases. Strings made of 5 to18 carbon atoms are liquids, and strings with 19 or more carbon atoms are solids.

So where do these names (methane, ethane, propane, butane) come from? What do they mean? An organization called the **International Union of Pure and Applied Chemistry (IUPAC)** decides what to name molecules and chemical compounds. Chemists all over the world need to use the same names for things so that they can discuss their work with each other. If a chemist speaks about "methanol" or "ethylene glycol," all the other chemists need to know exactly what substance is being discussed. Sometimes, IUPAC decides to go with names that chemists have already been using for a while. Other times, IUPAC decides to change the name to something more logical. The goal is to have a naming system with rules that everyone knows, so that there is as little confusion as possible. And confusion is a distinct possibility in a science where there are millions of molecules that could be named.

The first step in naming a carbon compound is to count how many carbon atoms are in it. This is how you count carbons:

| 1 | 2 | 3 | 4 | 5 | 6 | 7 | 8 | 9 | 10 |
|---|---|---|---|---|---|---|---|---|---|
| meth- | eth- | prop- | but- | pent- | hex- | hept- | oct- | non- | dec- |
| | | *(prope)* | *(byute)* | | | | | *(known)* | *(deck)* |

These are the prefixes that come before suffixes like "ane" or "ene" or "yne." In this chapter we are talking about alk*anes*, so each of these prefixes has "ane" after it. "Ane" simply means single-bonded carbons. We have seen methane, ethane, propane, and butane. We can now add pentane, hexane, heptane, octane, nonane, and decane.

You might recognize the word **octane**. This word is found on gas pumps where they post "octane ratings." A gasoline that has an octane rating of 87 means that the gasoline is 87% octane and 13% heptane. Inside the engine, the fuels get compressed before they are ignited by the spark plug. Heptane has the unfortunate characteristic of exploding too early, before it is ignited by the spark. This causes something called "knocking" in the engine, which is not desirable. Octane can handle the compression much better. So, the more octane, the better. Unfortunately, the higher the octane rating, the higher the price, also! Better things always cost more, don't they?

Yikes! Alkanes are expensive!

REGULAR ALKANES $2.09

Chains of 12 to 16 carbons give you kerosene fuels.  15 to 18 carbons make heating oil.  20 to 40 carbons give you paraffin waxes and asphalt. Strings of hundreds or thousands of carbons make various kinds of plastics. (Plastics will have their own chapter.)

| Number of carbons | Uses |
|---|---|
| 1-4 | natural gas (used for fuel) |
| 5-12 | gasoline, solvents |
| 13-16 | kerosene, diesel fuel, jet fuel, heating oil |
| 17-20 | lubricating oils |
| 21+ | paraffin, asphalt |

All the products listed above can be made from the same raw material: **crude oil**.  "Crude" just means raw or unrefined—the natural stuff as it comes up from the ground.  Scientists guess that crude oil was formed by the decomposition of plants and animals under great pressure a long time ago.  Crude oil is made of alkanes.

A factory called a **refinery** can sort out the different lengths of alkanes in crude oil.  The refinery uses a process called **distillation**.  You may be thinking of distilled water and wondering if there is a connection.  Yes, the process of distillation is similar no matter what you are distilling.  **Distilling** means heating a substance until it turns to steam, then gradually cooling it.  As it cools, it turns into a liquid.

Crude oil is heated until it turns into vapor (at 350° C), then this vapor is pumped into the bottom of a very tall tube.  The temperature is hot at the bottom, and cooler at the top.  The longest hydrocarbon chains turn back into a liquid (condense) onto trays at the bottom of the tube and run into pipes.  The next-longest hydrocarbons liquefy at the next level up and run into those pipes, and so on, until the very shortest hydrocarbon chains, such as methane and propane, are collected at the top.

You can see some videos about this on the Carbon Chemistry playlist at www. YouTube.com/The BasementWorkshop.

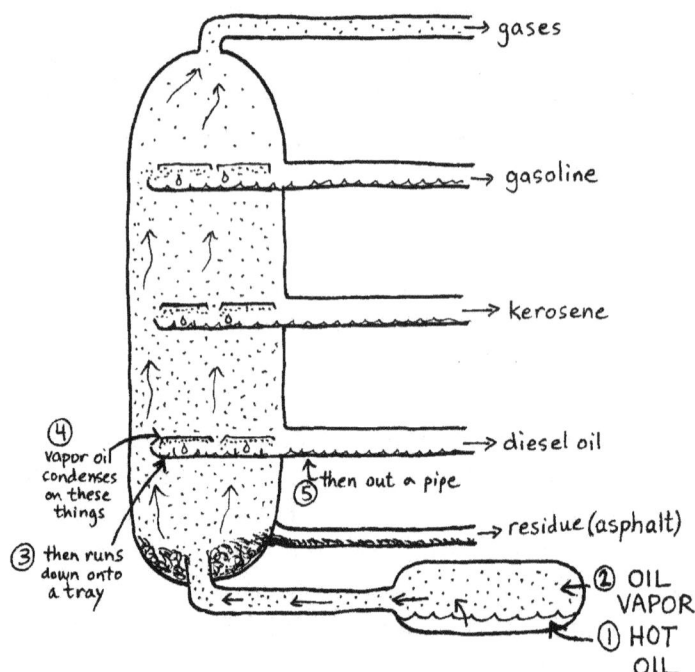

**Cracking** is a process by which they take medium-sized chains and break them into smaller pieces.  The medium-sized chains are heated in the absence of oxygen, and sometimes in the presence of chemicals called catalysts, which help the reaction occur. A commonly used catalyst is "zeolite," a mineral powder made of aluminum, silicon and oxygen.  Cracking can produce liquid gasoline (petrol), a substance which is always in demand.

Two more ideas that we need to discuss in this chapter are chlorinated hydrocarbons and isomers.  Let's tackle **chlorinated hydrocarbons** first.

As we mentioned in chapter one, carbon can bond with almost any atom that is willing to share an electron. Hydrogen very often does this, but other atoms do, too. Chlorine is an atom that has only seven electrons in its outer shell—three happy pairs and one very unhappy electron that is all alone. Chlorine gladly attaches itself to carbon. Here are four examples of molecules where one or more hydrogens are replaced by chlorine:

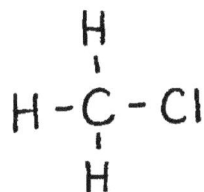

$$H-\overset{\displaystyle H}{\underset{\displaystyle H}{C}}-Cl \qquad H-\overset{\displaystyle H}{\underset{\displaystyle Cl}{C}}-Cl \qquad Cl-\overset{\displaystyle H}{\underset{\displaystyle Cl}{C}}-Cl \qquad Cl-\overset{\displaystyle Cl}{\underset{\displaystyle Cl}{C}}-Cl$$

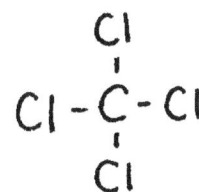

methyl chloride        methylene chloride        chloroform        carbon tetrachloride

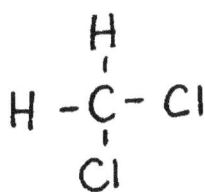

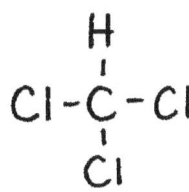

Methyl chloride is mainly used in making silicone substances (sealants, waterproofing materials, artificial body parts, Silly Putty). Methylene chloride is used as a paint remover. Chloroform started out as an anesthetic (putting you to sleep for surgery), but has now been replaced by safer substances. Chloroform is sometimes referred to as "knock-out gas." Bad guys in movies soak handkerchiefs in chloroform and put them over the faces of their victims. Carbon tetrachloride was formerly used in dry-cleaning, but is no longer used because of safety concerns. It can react with water to produce a poisonous gas.

These substances do not dissolve in water, which is a problem when they escape out into nature. **DDT** (di-chloro-di-phenyl-tri-chloro-ethane) is famous for both its effectiveness as an insecticide (killing insects) and, unfortunately, its ability to harm wildlife such as birds and reptiles, mainly through the birth of deformed babies. DDT was so effective at killing moquitoes, that malaria, a disease carried by mosquitoes, was eliminated in North America. Africa now faces the problem of trying to control malaria without DDT.

Carbon compounds can contain fluorine along with chlorine. The fluorine atom is in exactly the same state as the chlorine atom, with one unhappy, unpaired electron. Fluorine will gladly attach itself to one of carbon's electrons.

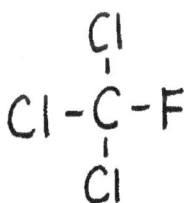

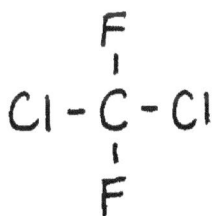

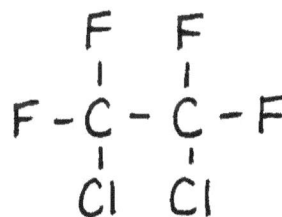

$$Cl-\overset{\displaystyle Cl}{\underset{\displaystyle Cl}{C}}-F \qquad Cl-\overset{\displaystyle F}{\underset{\displaystyle F}{C}}-Cl \qquad F-\overset{\displaystyle F}{\underset{\displaystyle Cl}{C}}-\overset{\displaystyle F}{\underset{\displaystyle Cl}{C}}-F$$

These molecules are called (no big surprise here) **chlorofluorocarbons**, or CFCs. They used to be used as propellants in aerosol spray cans and as coolants in refrigerators. CFCs are nontoxic and don't hurt us directly because they don't react chemically with anything. They don't pose any direct health hazards to humans. The problem with them is that when they are released into the air, they float up into the atmosphere where they are changed by ultraviolet light into molecules that can damage the protective ozone layer of the atmosphere. In order to protect the ozone layer, most governments have banned the use of CFCs. In some cases, CFCs have been replaced with HFCs, hydrofluorocarbons.

He's had it.
I'll turn the page with you.

No more!
My brain is full!

One last topic remains: *isomers*. The name sounds strange, but the idea is very easy. Isomers are molecules with exactly the same number of atoms but in a different geometrical arrangement. ("Iso" means "same.")

For example, let's look at butane, $C_4H_{10}$. The most obvious way to arrange the atoms is like this:

$$
\begin{array}{cccc}
H & H & H & H \\
| & | & | & | \\
H - C - C - C - C - H \\
| & | & | & | \\
H & H & H & H
\end{array}
$$

However, you could reshuffle the carbons a bit and make the molecule look like this:

$$
\begin{array}{ccc}
H & H & H \\
| & | & | \\
H - C - C - C - H \\
| & | & | \\
H & & H \\
& H - C - H \\
& | \\
& H
\end{array}
$$

It is still $C_4H_{10}$, butane. To differentiate it from regular butane, scientists call this "isobutane," an isomer of butane.

Here are three isomers of pentane, $C_5H_{12}$.

$$
\begin{array}{ccccc}
H & H & H & H & H \\
| & | & | & | & | \\
H - C - C - C - C - C - H \\
| & | & | & | & | \\
H & H & H & H & H
\end{array}
$$

Why are isomers worth mentioning? One practical use for isomers is in gasoline. Petroleum chemists have found that branched isomers of octane actually burn better than straight octane. Branched nonane and decane are also put into gasoline. The chemists alter the straight alkanes that come from the refinery, adding chemicals that cause them to rearrange into branched isomers.

We didn't draw all the H's, but you know they are there, right?

Sad, isn't it?

Zzzzz

## Comprehension self-check

      See if you can fill in the blanks and answer these questions, based on what you remember reading.  If you have trouble, go back and re-read.

1) The simplest hydrocarbon is called _____.
2) Methane is also called _____ gas.
3) Does methane burn easily? ____
4) Is methane smelly? _____
5) Where can methane be found in our bodies? _____
6) Methane, ethane, propane, etc. belong to a group of molecules called _____s.
7) What does the IUPAC do? _____
8) Can you put these prefixes in order, from 1 to 6?  (but, meth, prop, hex, pent, eth)

_____, _____, _____, _____, _____, _____
9) Where can you find octane and heptane mixed together? _____
10) Too much heptane and not enough octane causes this problem: _____
11) Use these words to fill in the blanks. [solids, liquids, gases]
Very short alkanes are _____, medium sized are _____ and longer ones are _____.
12) Raw or unrefined oil is called _____ oil.
13) A factory that processes oil is called a _____.
14) The primary method factories use to refine oil is called _____, which is heating the oil, then allowing it to cool and condense.
15) Breaking hydrocarbon chains into smaller pieces is called _____.
16) Name two other atoms, in addition to hydrogen, that will bond with carbon: _____ and _____.
17) CFC's contain these three types of atoms: _____, _____ and _____.
18) DDT was used to kill _____ but it also killed _____.
19) What was chloroform first used for? _____
20) Molecules that contain exactly the same number of atoms, but in a different geometric arrangement, are called _____.

Write the matching letter on each line.

21) methane ____   22) ethane ____   23) propane ____   24) butane ____   25) octane ____

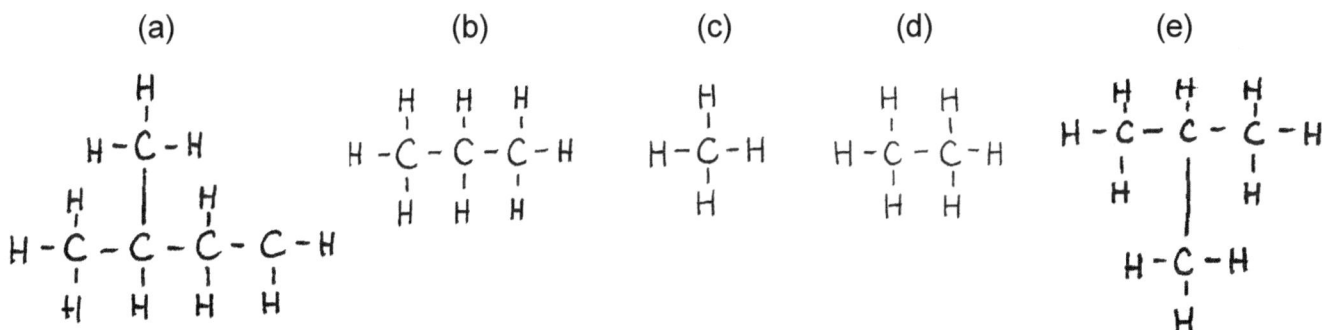

   (a)               (b)             (c)           (d)           (e)

## Hydrocarbon puzzle

Here are the clues for the missing words.   You have to figure out which goes where!

- A factory where hydrocarbons are processed
- The primary method factories use to process hydrocarbons
- Gasoline is mostly this hydrocarbon
- This hydrocarbon is found in handheld lighters
- This hydrocarbon is found in gas grill tanks
- This atom is the "F" in CFC's

- This is the word for hydrocarbon chains with only single bonds
- Hydrocarbons burn easily. They are highly _____.
- Chlorofluorocarbons destroy the _____ layer of the atmosphere.
- This method is used to break apart hydrocarbon chains.
- This hydrocarbon is natural gas.

H
Y
D
R
O
C
A
R
B
O
N

## "Cross one out" puzzle

1) Which one of these is not a hydrocarbon?
   methane    gasoline    diesel fuel    rubbing alcohol    kerosene    asphalt    ethane

2) Which one of these has nothing to do with refining crude oil?
   distillation    fermentation    condensation    evaporation

3) Which one of these is not an alkane?
   propane    butane    ethyne    nonane    decane    heptane    methane    octane

4) In which one of these places are you least likely to find natural gas?
   mountain tops    grill tanks    intestines    swamps    silos    mines

5) Which one of these is not an alkane hydrocarbon?
   $CH_4$        $C_4H_{10}$        $C_3H_8$        $C_2H_2$        $C_5H_{12}$

6) Which one of these does not bond with carbon?
   chlorine    fluorine    carbon    hydrogen    helium

7) Which of these is not considered to be a way of modeling an atom?
   empirical formula    structural formula    space-filling    ball and stick

8) Which of these would not come out of an oil refinery?
   kerosene    gasoline    CFCs    methane    asphalt    diesel oil

# CHAPTER 3: "-ENES" AND "-YNES"

In chapter one, we mentioned that carbon can sometime form what we call a "double bond." Then in chapter two, we saw nothing but single bonds. Alkanes molecule have only single bonds. Now we are going to look at some molecules with double, even triple, bonds.

What would happen if we plucked some hydrogens off an alkane? Let's try it.

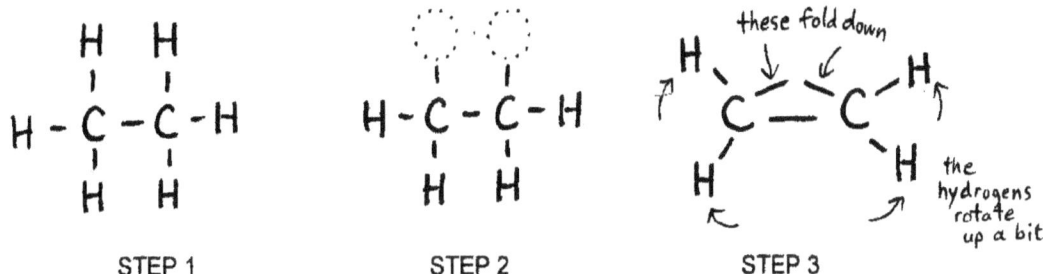

| STEP 1 | STEP 2 | STEP 3 |

Look at what happened—the carbons tilted themselves a bit so that their unpaired electrons matched up with each other. That seems to work out pretty well. However, we no longer have an alkane, we have an **alkene**. Alkenes are molecules similar to alknanes, except that somewhere in the molecule there are some carbons forming a double bond with each other.

The alkene shown here, in step three, is called **ethylene**. The name "ethylene" might sound like a poison, but it is actually a harmless gas produced naturally by ripening fruit. es, that bowl of fruit there on the table is giving off ethylene gas. Commercial produce growers have found that they can speed up the ripening process by steaming their produce in ethylene. Tomatoes, especially, can be "reddened up" by exposing them to ethylene gas. Unfortunately, the taste does not improve as fast as the color does. "Gassed" winter produce may look good, but it doesn't taste like that vine-ripened summer stuff. Ethylene gas is also used as an ingredient in automobile anti-freeze (ethylene glycol) and in plastics called polyethylene. We will learn more about these plastics in a later chapter.

Can chemists change ethene back into an ethane by adding some hydrogens? Yes, they can. Here is how they write this process:

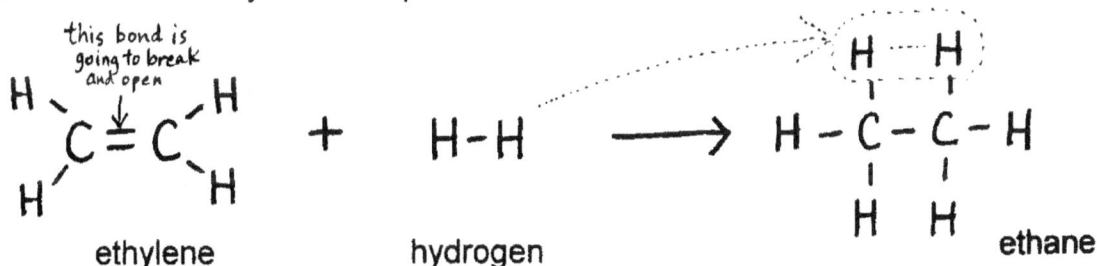

ethylene        hydrogen        ethane

Alkane molecules are **saturated** molecules. Saturated means completely full. When a sponge is saturated, it can't hold any more water. When a liquid is saturated, it can't hold any more of whatever you are trying to stir into it. When your brain is saturated (like our poor little friend's) it can't hold any more information and you need to have a recess or take a nap. When a hydrocarbon is saturated, it has all the hydrogen atoms it can possibly hold on to. The opposite of saturated is **unsaturated**. If a hydrocarbon has some double bonds that it could open up, then it is unsaturated.

SATURATED                                    UNSATURATED

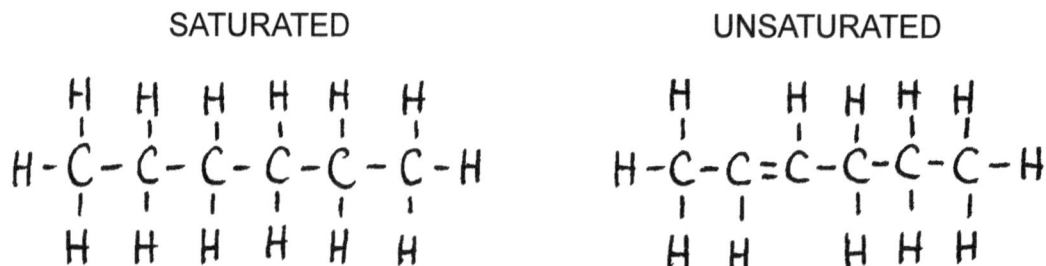

Now, let's look at a molecule with a triple bond. If we pluck off another pair of hydrogens, carbon has no choice but to increase the double bond to a triple bond, sharing not two, but three pairs of electrons. Carbon must feel like a contortionist from an old-fashioned circus! (Contortionists are performers who can bend their bodies in all sorts of unnatural ways.)

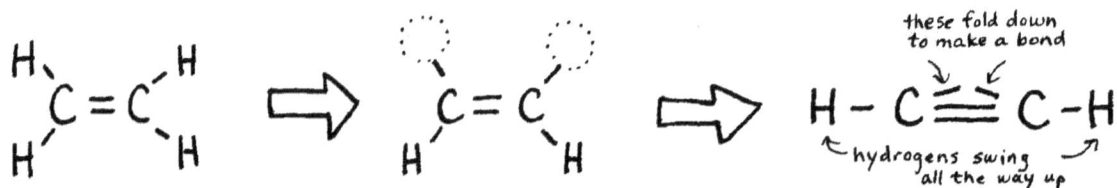

This molecule is the smallest member of the **alkyne** group. Alkynes all have at least one carbon with a triple bond. To really confuse you, the name of this chemical is "acetylene," the fuel found in acetylene torches. (Hmm… maybe IUPAC didn't get there in time and acetylene got named by someone not familiar with the ane/ene/yne naming rule.)

Alkenes and alkynes have many of the same properties as alkanes. Molecules with only 2 to 4 carbons are gases, 5 to 18 carbons are liquids, and 19 or more carbons are solids. They do not dissolve in water; they float on water. (Think of an oil spill.)

Another group of molecules that ends in "-ene" is the **benzene** group. This group is also called the **aromatic hydrocarbons** because the first ones that were discovered had strong aromas (smells). Later on, scientists discovered that not all members of this group have odors, but it was too late—the name had been used for so long that it was impossible to change it.

The discovery of the structure of benzene has a famous story attached to it. Benzene's empirical formula, $C_6H_6$, had been discovered in 1825 by Michael Faraday, a scientist much better known for his work with electricity and motors. Once the ration of carbons to hydrogens was known, many chemists drew up every possible configuration they could think of for $C_6H_6$. Here are two of them:

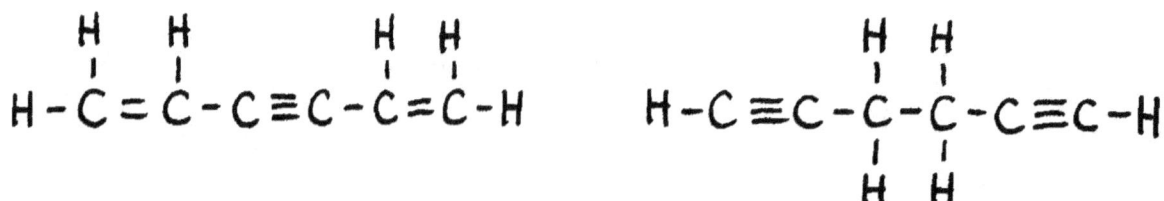

All of the configurations contained several double and/or triple bonds. However, chemistry experiments with benzene clearly demonstrated that benzene couldn't have double and triple bonds. What was up? No one could figure it out for 40 years. Then along came a scientist named August Kekule *(keck-u-lay)*. He worked on the puzzle long and hard until it almost drove him crazy. One night he fell asleep thinking about the benzene puzzle. As he slept he dreamed that the straight molecule curved around until its ends touched, forming a ring. (Some versions of the story say that he saw a snake biting its tail.) When he woke up he realized he had solved the puzzle in his sleep. Benzene must be a ring!

There was more work to do after he woke up, however. This structure still had a bonding problem. If you look at the molecule in his dream bubble and count the bonds attached to each carbon, you will see that there are only three. Carbon atoms always make four bonds. What was happening? This solution was proposed:

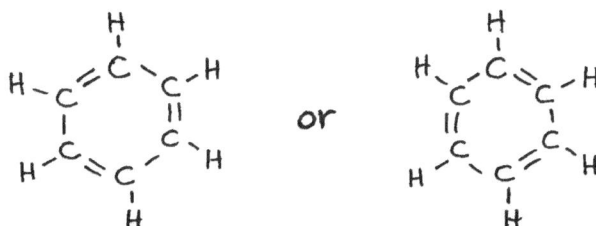

The bonds between the carbons alternate back and forth between single and double so fast that they are neither single nor double. The electrons are not tied down to any one place, but are spread around the ring. After deducing that this must be the case, chemists decided to draw the benzene structure like this:

The vertices ("corners") of the hexagon represent the six carbon atoms. There are H atoms at each vertex, but they are rarely drawn.

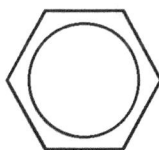

The circle represents the electrons that are being shared back and forth in the single and double bonds.

The benzene ring is a very stable structure. Stable molecules are generally less dangerous than unstable ones. If it were not for your liver, benzene would not be considered poisonous. Benzene would just float through your body, not bothering any of your cells. But when benzene arrives in the liver, the liver starts to disassemble it. The liver is just doing its job— it's supposed to take chemicals apart and get rid of them. This time, however, disassembly is not helpful. But your liver doesn't know this, and it starts messing with benzene, disturbing the stable structure. Now, thanks to your liver, you have an unstable, dangerous substance in your body. Your bone marrow is the body part most affected, and it begins to produce fewer and fewer red blood cells, leading to a condition called anemia, where you feel very tired because you are not getting enough oxygen. Prolonged exposure can also lead to cancers of white blood cells made in the bone marrow.

If you add a $CH_3$ to one edge of the benzene ring, you get ***toluene*** *(TALL-yu-een)*, a very useful chemical,  but a dangerous one.  Toluene is more harmful than benzene.  Often, a very small change in a molecule can result in very large changes in its behavior.

Don't forget that there are H atoms at the other five vertices ("corners").  Chemists don't draw them because they assume everyone knows that they're there.

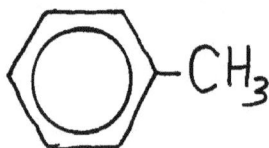

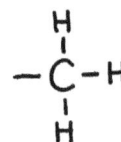

We could draw $CH_3$ like this.  It is usually written "$CH_3$" just to save space.

Toluene is a clear, fragrant liquid which is used as a solvent in products such as paints, varnishes, cleaning products, pesticides, adhesives and explosives.  You almost certainly have products in your home that contain toluene. Toluene is a hazardous chemical and it must not be dumped down drains. It has been proven to cause cancer in laboratory animals. To get rid of it, you must take it to an official hazardous waste collection site.  Fortunately, toluene does eventually break down, so it won't stay in the environment forever.

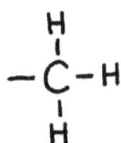

$CH_3$ is often referred to as the "methyl" group.

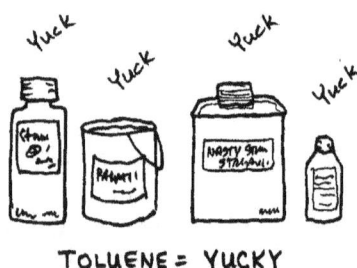

Yuck  Yuck  Yuck  Yuck

TOLUENE = YUCKY

Yup.   Hey- we're on the last page already!

One interesting side note about toluene is that it is produced naturally by the tolu tree. This tree grows in South America and is tapped (like a maple tree) to get the tolu resin, which is used many food and health products. The native peoples use tolu resin to make medicines.

If you take a toluene molecule and add another $CH_3$ to it, you get **xylene** *(ZIE-leen)*. Xylene was first discovered by a French chemist in 1850 as he was pulling chemicals out of wood tar. Since he found it in wood, he used the Greek word for wood, "xylon," to name it.  Xylene was found to be useful in labs as a solvent and a cleaning agent because it could dissolve substances like wax and tar.  The printing and rubber-making industries soon found uses for xylene. Today, millions of tons of xylene are produced every year for the manufacuring of plastics, especially the type used for plastic bottles and polyester fabrics.

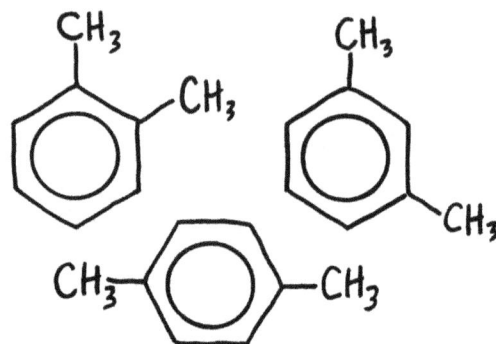

Xylene has three isomers.  The $CH_3$ can be in any of these positions.

If you join two benzene rings together, you get a molecule of ***naphthalene*** *(NAFF-thall-een)*. You know naphthalene as moth balls.  Moth balls release a gas that is toxic to the larvae of moths that like to eat clothing made of natural fibers like wool. It's easy to remember that this molecule is the one that kills moths because the molecule actually looks a moth.

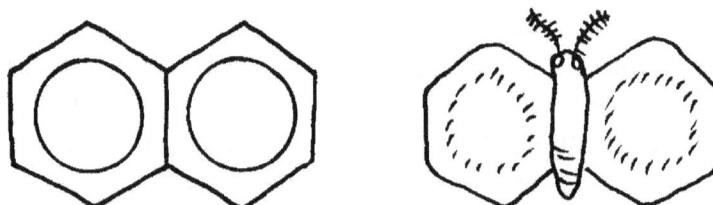

## Comprehension self-check

See if you can fill in the blanks and answer these questions, based on what you remember reading.  If you have trouble, go back and re-read.

1) Alkanes have _____ bonds, alkenes have _____ bonds and alkynes have _____ bonds.

2) Hydrocarbons that contain just 2-3 carbon atoms are g_____.  Large molecules with more than a dozen carbon atoms are s_____.

3) Ethylene gas is produced naturally by _____.

4) Hydrocarbons that have the maximum number of hydrogens they can possible hold are said to be s_____.

5) Hydrocarbons that have double or triple bonds that could be opened up (to allow for the addition of more hydrogens) are said to be u_____.

6) How many carbon atoms are in a benzene ring? _____

7) About how many years did it take for scienetists to figure out the shape of benzene? _____

8) Benzene rings have strange bonds that alternate between _____ and _____.

9) Molecules that contain at least one benzene ring are called _____ hydrocarbons because the first examples discovered were smelly substances.

10) Benzene becomes harmful to us when our _____ starts to disassemble it.

11) Which is more harmful, benzene or toluene? _____

12) How many benzene rings does toluene contain? _____

13) How many benzene rings does xylene contain? _____

14) Most xylene produced today is used to make _____ and

_____.

15) How many benzene rings does naphthalene contain? _____

16) What household product is made of naphthalene? _____

17) What does the naphthalene molecule resemble? _____

18) True or false?  It's okay to dump toluene down the drain. _____

19) Review:  What do you call molecules that have the same empirical formula (such as Cbut differ in their shape? _____

20) Review:  $CH_4$ is known as _____ or _____.

The end of the chapter!

## Draw the bonds

Put all the bond lines between the letters.  Make sure that each carbon, "C," has four bond lines going out from it.  H's can only have one line sticking out from them and they never attach to each other, only to carbons.  A carbon can have more than one H attached to it.

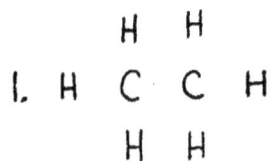

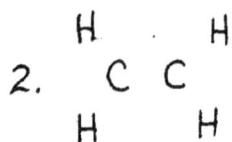

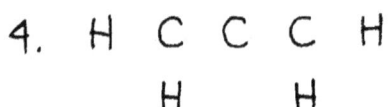

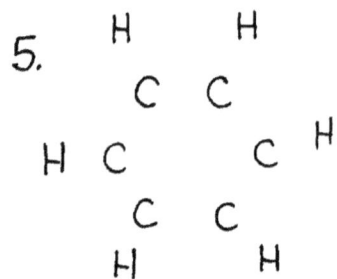

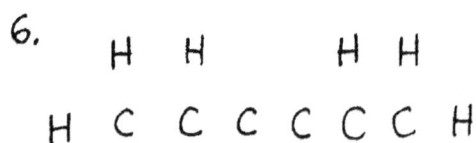

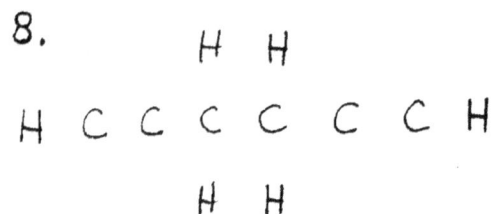

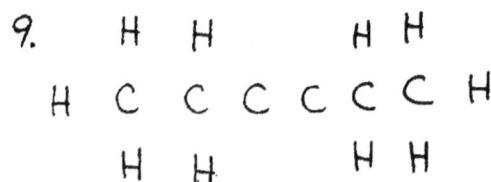

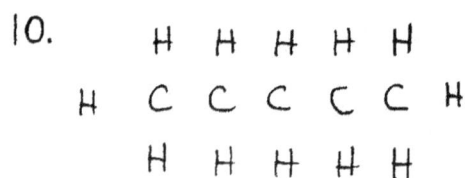

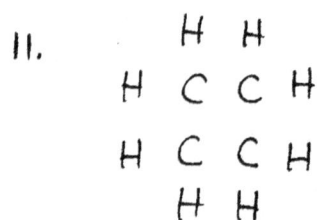

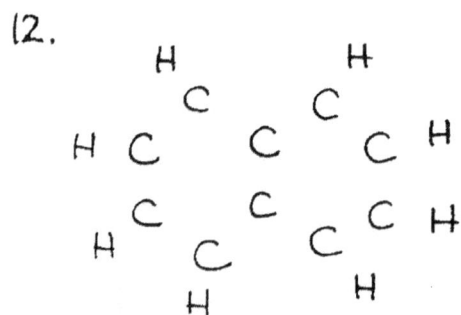

```
      H   H
1. H  C   C  H
      H   H
```

```
    H       H
2.    C   C
    H       H
```

```
3.  H  C   C  H
```

```
4.  H  C   C   C  H
          H       H
```

```
5.    H        H
        C   C
    H C         C  H
        C   C
      H        H
```

```
6.  H  H        H  H
    H  C  C  C  C  C  C  H
```

```
7.  O  C  O
```

```
8.        H  H
    H  C  C  C  C  C  C  H
              H  H
```

```
9.  H  H        H  H
    H  C  C  C  C  C  C  H
        H  H        H  H
```

```
10.    H  H  H  H  H
    H  C  C  C  C  C  H
       H  H  H  H  H
```

```
11.      H  H
       H  C  C  H
       H  C  C  H
          H  H
```

```
12.    H            H
         C        C
    H C      C      C  H
       C        C
    H       C       C  H
         C      C
       H            H
```

The carbons will form two rings that are attached in the middle.

# CHAPTER 4: FUNCTIONAL GROUPS

Building organic molecules is a little like building with a construction toy. You can pop pieces on and off, or add extra parts. If you take the end hydrogen off an alkane, you could stick on something else.

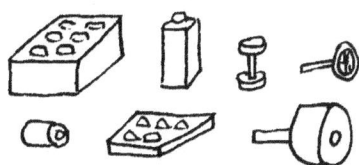

child's building toy

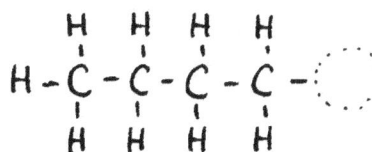

chemist's building toy!

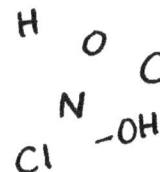

What could you put on? If it were a toy, you might add a red brick or a set of wheels. But since it is a molecule, your only options are things like oxygens, nitrogens, or more hydrogens. Here is a sample of some of the parts we can stick on an alkane:

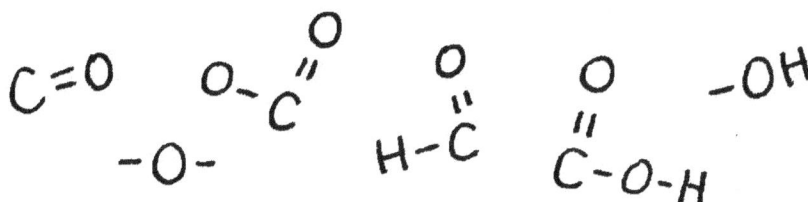

They're boring and they all look the same!

Each of these parts can turn the alkane into something totally different, just like adding wheels to a Lego block can turn it into a vehicle. Chemists don't call these things "parts," they call them **functional groups**, because each one will make a molecule function a certain way, just like those wheels make the block function as a vehicle. These functional groups make the molecules function as chemical things like alcohols or acids.

Here are the groups we are going to talk about in the rest of the chapter:

<u>Name of functional group</u>　　　　　<u>What it looks like</u>

Alcohol  (-OH)

$-OH$

Carboxyl  (-COOH)

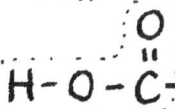

Aldehyde  (-CHO)

I'd rather build with Legos!™

Ketone  (-CO)

$C=O$

Ester  (-COO-)

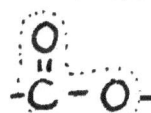

Ether  (-O-)

$-O-$

23

Let's look at each group separately, starting with alcohol.

We will take the smallest alkane, methane, pop one hydrogen off, and stick on the functional group "OH."

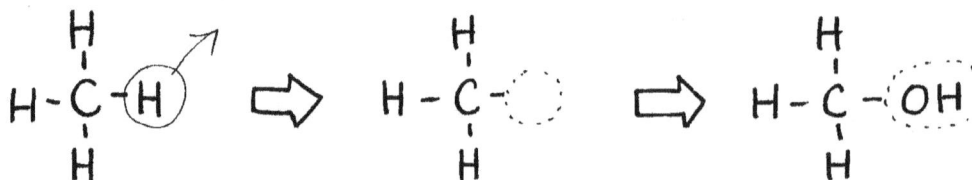

This type of alcohol is called **methanol**. "Meth-" means "one," "an" reminds you it came from an alk<u>an</u>e, and "-ol" means it is an alcohol. This substance actually has several names because the IUPAC didn't yet exist when chemists began discussing methanol. Some chemists called it "methyl alcohol," and others called it "wood alcohol" because you can make it from wood, like this:

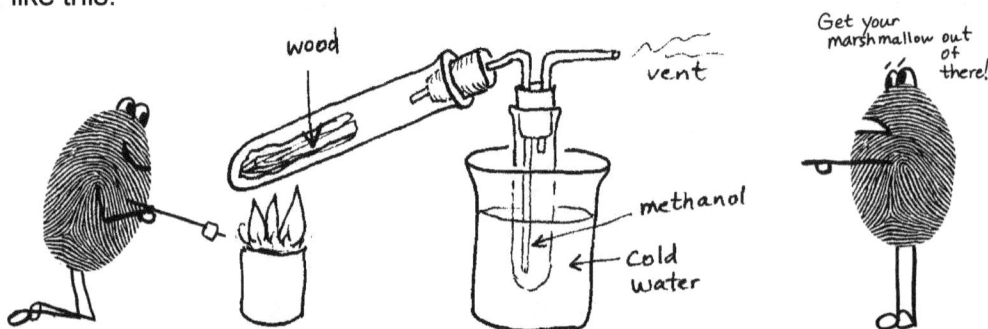

Methanol is poisonous. It is used in industry as a solvent and as an ingredient for more complex chemicals. Researchers are experimenting with it to see if it could be used as a fuel for vehicles.

Now let's take the next-to-smallest alkane, ethane, and stick the "OH" on the end.

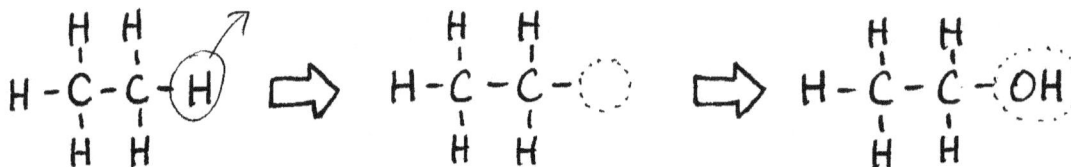

What would this be called? If methane turned into methanol, then ethane would turn into... **ethanol**, right? Right! However, (here comes another naming problem) chemists used to call it ethyl alcohol, and you will still hear this term used. And, just like with methanol, there is a third name. Ethanol is sometimes called "grain alcohol" because it is commonly made from the fermentation of grains or other starchy plants.

Ethanol is the alcohol in beverages such as beer and wine. Like all alcohols, ethanol is technically a poison. Consumed in small amounts, the alcohol doesn't kill, it just slows down the brain and nerves. People interpret a mild slowing down of the nervous system as enjoyable. After a certain point, however, the slowing down of the nervous system becomes dangerous, especially if the affected person is driving a car. Consuming large amounts of ethanol too quickly can cause unconsciousness, or even death. Beverages that have been "distilled," such as whiskey and vodka, have much high concentrations of ethanol and are therefore much more dangerous than wine or beer.

Let's look at one more alcohol. We'll use propane as our base:

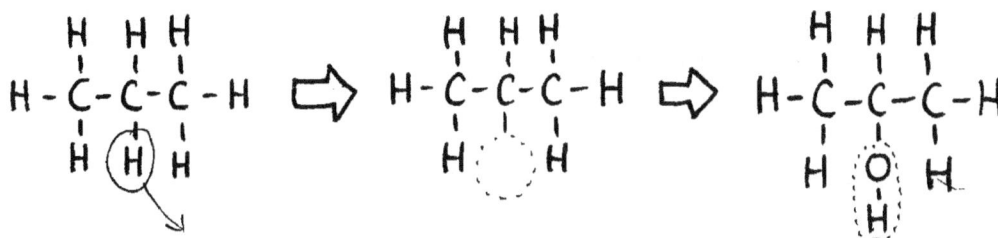

H - C - C - C - H  ⟹  H - C - C - C - H  ⟹  H - C - C - C - H

Since it is sticking off the middle carbon it's an isomer of propanol, thus it is isopropanol.

The names for this substance are: isopropanol, *isopropyl alcohol*, and rubbing alcohol. You probably have a bottle of this in your medicine cabinet at home. It is very poisonous, which makes it excellent as a disinfectant. Nurses always rub isopropanol on your arm before giving you a shot so that the needle doesn't push germs under your skin.

Let's try another functional group. We will pop an H off a methane and stick on a COOH.

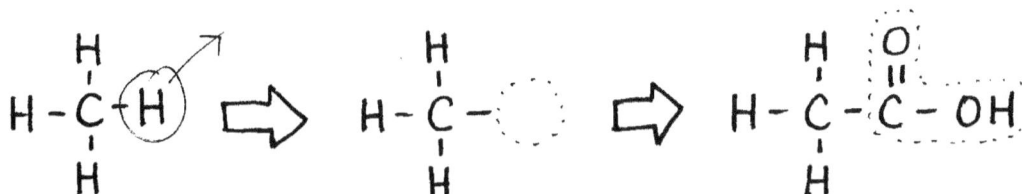

H - C - H  ⟹  H - C -  ⟹  H - C - C - OH

We have made an acid called **acetic acid**, which is found in vinegar. The Latin word for vinegar is "acetum." **Carboxylic acids** have strong, pungent odors. Oddly enough, that H on the end of the COOH usually goes wandering off, leaving just COO-. Since a hydrogen atom is nothing but a proton and an electron, it can leave its electron with the molecule and go off as nothing but a proton. The definintion of an acid is a substance with lots of free protons.

If you add COOH to butane and turn it into a carboxylic acid, you get **butyric acid**, which is one of the foulest smelling substances there is. Rancid (rotten) butter contains butyric acid. Human body odor is partly caused by small amounts of butyric acid.

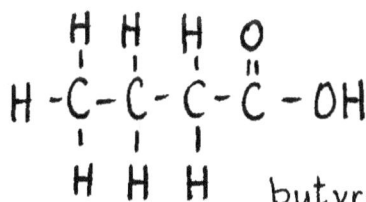

H - C - C - C - C - OH

butyric acid (byu-teer'-ick)

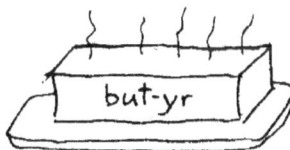

but-yr

butter
⇕
butyr

A new reason to hate chemistry!

If you stick the COOH on just a hydrogen, you get **formic acid**, which is produced by ants. (The Latin word for ant is "formica.")

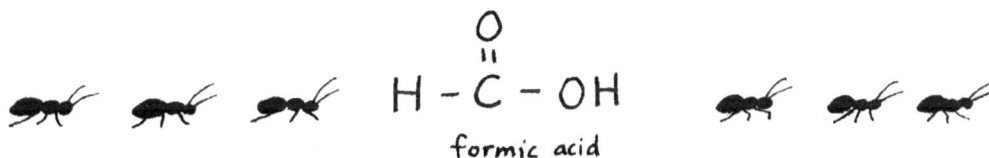

O
‖
H – C – OH

formic acid

The sting you feel when an ant bites you is not from its teeth, but rather from the formic acid it injects into your skin.

The other carboxylic acids aren't things you are familiar with, so let's save your brain space for the other functional groups.

Let's do **aldehydes** and **ketones** next. If you stick the aldehyde group (-CHO) on a methane, you get a substance known as **acetaldehyde**.

We didn't show the ripping off of the hydrogen this time.

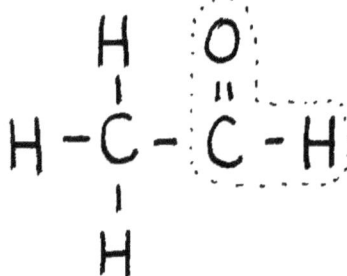

We figured you probably had it down pretty well but we still put the dots around the functional group.

Never heard of it, right? Okay, let's do one you might be familiar with. Let's put the aldehyde group (-CHO) on just a hydrogen.

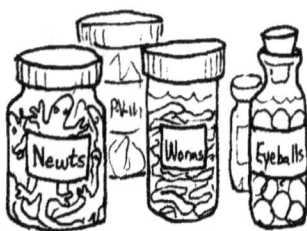

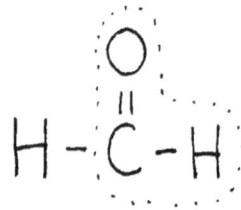

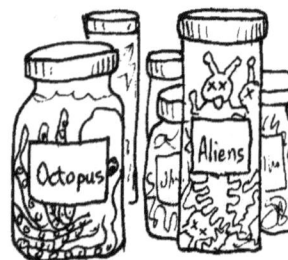

This is **formaldehyde**. It is a gas at room temperature, but can easily be dissolved into water to make **formalin**, a liquid used in labs to preserve biological specimens in jars. Formaldehyde is used in industry to make plastics and to disinfect buildings.

**Ketones** have the **carbonyl** (C=O) functional group. The smallest ketone is **acetone**.

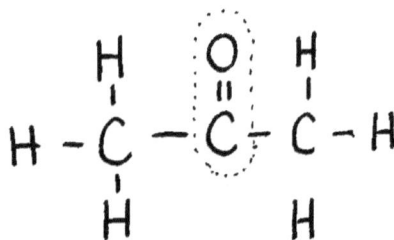

It's hard to believe those letters are actually dangerous!

Acetone is used by manufacturers of rubber, plastics, and varnishes. It is used as a varnish remover because it can dissolve dried-on varnish. One type of varnish it can remove is fingernail polish, so acetone is often the primary ingredient in fingernail polish removers. (Due to safety concerns many polish removers now use a less toxic replacement for acetone.) Bug collectors have been known to kill bugs by putting them in a jar with a cotton ball that was soaked in acetone. Fortunately, that much acetone won't kill us. The other ketones are also solvents used in industry. We won't bother you with their names or chemical formulas.

We've done enough stinky things; let's talk about some sweet smells. Let's look at some **esters**. The esters have pleasant, fruity smells, which is ironic because the way esters are manufactured is to combine two stinky things: a carboxylic acid and an alcohol. Two foul-smelling molecules can be combined to make some of the most delicious smells on earth.

Here is a chart showing some esters and what they smell like. The names of the esters come from the acid and the alcohol they were made from.

| Name | Empirical formula | What they smell like |
|---|---|---|
| Methyl butyrate | $CH_3CH_2CH_2COOCH_3$ | Apple |
| Ethyl butyrate | $CH_3CH_2CH_2COOCH_2CH_3$ | Pineapple |
| Propyl acetate | $CH_3COOCH_2CH_2CH_3$ | Pear |
| Pentyl acetate | $CH_3COOCH_2CH_2CH_2CH_3$ | Banana |
| Pentyl butyrate | $CH_3CH_2CH_2COOCH_2CH_2CH_2CH2CH_3$ | Apricot |
| Octyl acetate | $CH_3COOCH_2(CH_2)_6CH_3$ | Orange |
| Methyl benzoate | $C_6H_5COOCH_3$ | Kiwi |
| Ethyl formate | $HCOOCH_2CH_3$ | Rum |
| Benzyl acetate | $CH_3COOCH_2C_6H_5$ | Jasmine |

Don't the names sound horrible? If you saw these names listed as ingredients in your candy you might decide not to eat it! Speaking of horrible, our last functional group gets us back to bad smells: the **ethers**. Some wart removers contain ether. If you want to know what ethers smell like, take a whiff of Compound W™. Ethers are best known for the role they played in the history of medicine. Diethyl ether was once used to put people to sleep before surgery.

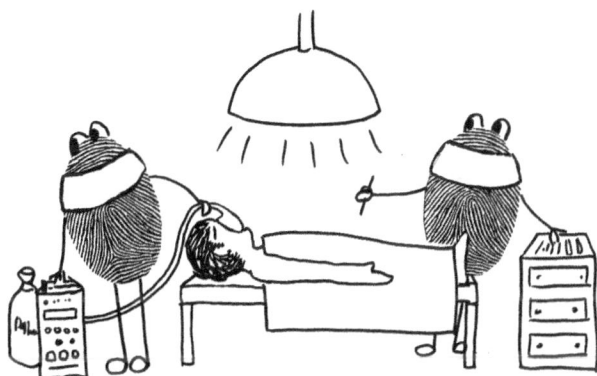

Ethers putting people to sleep during surgery.          Ethers putting people to sleep during class.

Diethyl ether caused side effects such as nausea and vomiting, and it has now been replaced by other substances, including other ethers. An ether that is important to the gasoline industry is known by its nickname, MTBE, because its real name is long and complicated. It is put into gasoline to reduce the amount of carbon monoxide in automobile exhaust.

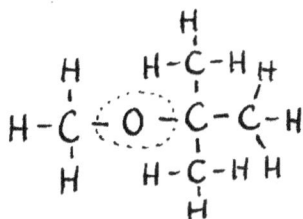

There are other functional groups, also. We didn't look at the phenols, the amines, or the amides. The amines show up when you study biochemistry and learn about proteins, which are made of amino acids. (Amino <u>acids</u> contain another functional group. Which one?) The next chapter will show us molecules that have more than one functional group.

## Comprehension self-check

See how well you can remember what you read.  Feel free to go back and look up the answers if you can't remember.

1)  Which one of these is NOT a functional group?

   a) alcohol        b) aldehyde     c) ketone     d) ethane     e) ether     f) ester

2) What does a functional group do?

   a) Makes a molecule poisonous.          b) Makes a molecule functional in a certain way.

   c) Makes a molecule smell bad.           d) Makes a molecule break apart.

3)  Adding an alcohol functional group (-OH) to methane turns it into _____, which can also be called _____ or _____.

4) Review: What does "meth" mean? _____

5)  Which type of alcohol is found in beer and wine? _____ Which alkane is this type of alcohol related to? _____

6) Review: What does "eth" mean? _____

7) What atoms do you add to a molecule to turn it into an alcohol? _____

8) What atoms do you add to a molecule to turn it into an acid? _____

9)  Name the carboxylic acid found in most kitchens: _____

10) Butyric acid gets its name from this food (when it spoils): _____

11)  What animal produces formic acid? _____

12)  What substance is used to preserve biological specimens? _____

13) Which is the smaller molecule, acetylaldehyde or formaldehyde? _____

14) What atoms do you add to a molecule to make it a ketone? _____

15)  What ketone is found in many nail polish removers? _____

16)  Which functional group makes molecules smell good? _____

17)  What does pentyl acetate smell like? _____ (You can look this one up.)

18)  What does ethyl butyrate smell like? _____ (You can look this one up, too.)

19)  Ethers were once used to _____.

20)  An ether named MTBE is put into _____ to reduce carbon monoxide.


## Review crossword puzzle for chapters 1-4

### ACROSS:

2) This is the smallest and simplest of the alcohols because it has only one carbon atom.

4) Molecules that have the same number and type of atoms but differ in shape are called ___.

5) Molecules that have benzene rings are called _____ because many of them have odors.

9) Diamond, graphite and coal are _____ of pure carbon.

10) An alkane hydrocarbon that has 8 carbons is called _____.

11) Ants produce this type of acid: _____

12) A hydrocarbon that has the maximum number of H's that it can hold is called _____.

13) The prefix "prop-" means _____.

15) The process a refinery uses to turn crude oil into usable products: _____

$$H-\overset{\overset{\displaystyle H}{|}}{C} = \overset{\overset{\displaystyle H}{|}}{C} - C \equiv C - \overset{\overset{\displaystyle H}{|}}{C} = \overset{\overset{\displaystyle H}{|}}{C} - H$$

16) "$C_3H_8$" is an example of a _____ forumula.

19) CFCs contain _____, fluorine and carbon.

21) A hydrocarbon that has at least one pair of triple-bonded carbon atoms is an ____.

22) The nucleus of an atom contains protons and _____.

23) If you add the alcohol functional group to ethane, you get _____.

24) Pencil "lead" is actually _____.   25) A single layer of graphite is called _____.

DOWN

1) The (COOH) functional group is called ____.          3) An alkene discovered in 1850 in wood tar.

4) The International Union of Pure and Applied Chemistry          6) Means "without shape"

7) Large hydrocarbon molecules are broken into smaller pieces using this process.

8) The type of chemical formula that uses letters attached to each other by lines representing bonds.

14) The smallest atom on the Periodic Table.          16) Artifical banana smell is an _____.

17) Rotten butter gets its smell from _____ acid.          18) An alkene that comes from the tolu tree.

20) An alkene gas produced by ripening fruit, and by chemical processes that make plastic.

21) A hydrocarbon that has at least one double bond but no triple bonds.

## MORE REVIEW!

Draw the bonds:

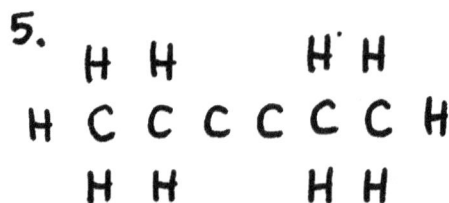

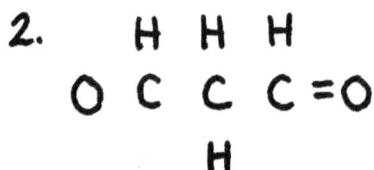

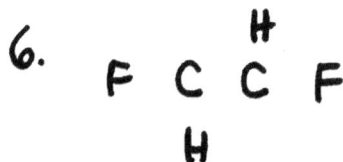

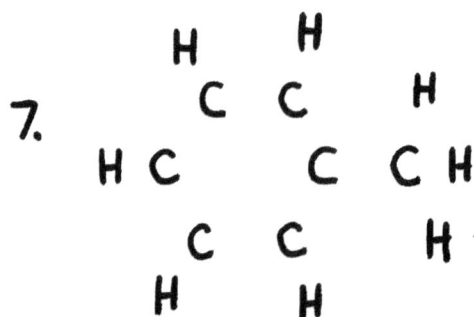

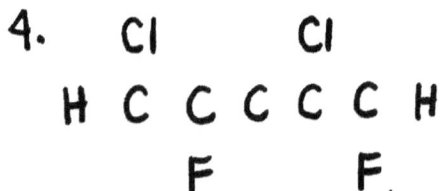

1.
```
        H
   H  C  C  C  H
      H  H  H
```

2.
```
      H  H  H
   O  C  C  C = O
         H
```

3.
```
      H           H
   H  C  C  C  C  C  H
```

4.
```
   Cl          Cl
   H  C  C  C  C  C  H
      F         F
```

5.
```
   H  H        H  H
   H  C  C  C  C  C  C  H
      H  H        H  H
```

6.
```
            H
   F  C  C  F
      H
```

7.
```
        H        H
        C     C        H
   H  C        C   C  H
        C     C        H
        H     H
```

8.
```
                     H
   H  C  C  C  C  H
      H        H  H
```

## Matching:

Match each molecule to the place you where you would be most likely to find it.

1) acetic acid ____          a. clothes closet
2) octane ____               b. car
3) graphite ____             c. gas barbecue grill
4) ester ____                d. kitchen cupboard (or frig)
5) formalin ____             e. French restaurant
6) ethylene ____             f. apple orchard
7) toluene ____              g. desk drawer
8) ethanol ____              h. painter's studio
9) propane ____              i. candy store
10) naphthalene ____         j. science lab

# CHAPTER 5: COMBINING FUNCTIONAL GROUPS

What would happen if you put more than one functional group on a molecule? What if you put a whole bunch of them all on one molecule? Would you get a jumbled mess, or would you get something useful? It could be either one, depending on what you did and how you did it. We can't possibly cover every single combination of functional groups, but let's look at some combinations that are well-known and very useful to chemists.

If you combine several carboxylic acids, you still have an acid, just a more complex one. Here are some examples of these more complicated acids.

*Wow! Now that's chemistry!!*

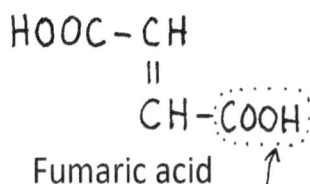

$$HOOC - CH$$
$$\parallel$$
$$CH - \overset{.}{\underline{COOH}}$$

Fumaric acid

*Remember, these things are carboxyl groups!*

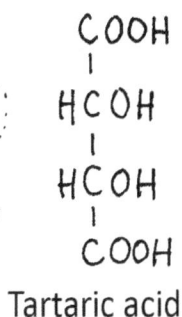

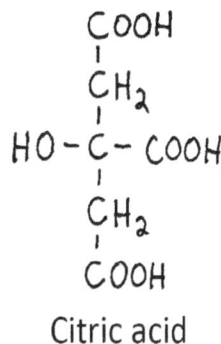

$$COOH$$
$$|$$
$$HCOH$$
$$|$$
$$HCOH$$
$$|$$
$$COOH$$

Tartaric acid

$$COOH$$
$$|$$
$$CH_2$$
$$|$$
$$HO - C - COOH$$
$$|$$
$$CH_2$$
$$|$$
$$COOH$$

Citric acid

*I know. That's why I have my eyes closed.*

All three of these acids are used in cooking and food processing. **Fumaric acid** and **tartaric acid** (cream of tartar) are used along with baking soda to make breads rise. **Citric acid** is used in food processing to add "tartness" to the flavor or to make the chemistry of the foods more acidic. There are lots more of them, similar to these. Many are used in making plastics or paints. But you really don't want to see more molecular structure drawings, so we'll move on.

What happens if you combine a carboxyl group (COOH) with a benzene ring?

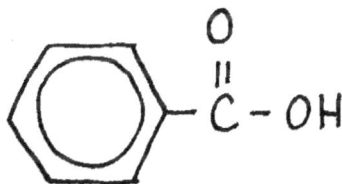

$$\overset{O}{\overset{\parallel}{C}} - OH$$

You get **benzoic acid**, an ingredient in **sodium benzoate**, a common food preservative. It stops mold and bacteria from growing in high-acid foods such as fruit juices, jams, salad dressings, pickles, and carbonated beverages. It's also used in some medicines and cosmetics. Look for "sodium benzoate" the next time you use a bottled food product.

If you add another carboxyl group, you get **phthalic acid** *(thall-ick)*, which is used in the production of paints and plastics.

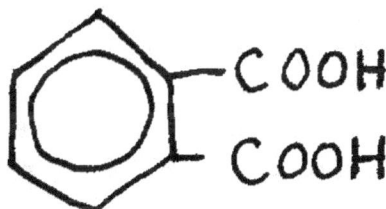

$$- COOH$$
$$- COOH$$

We could go on adding acids to benzenes... but we're not. You can study them later on, if you take organic chemistry in college.

Here's another idea: What if you put an aldehyde on a benzene ring? What do you get?

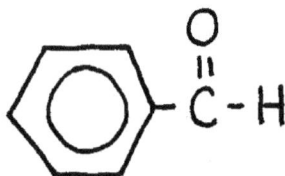

This is **benzaldehyde** (ben-ZALL-de-hide). It can be made naturally by cooking down almond and apricot pits, or artificially by adding chemicals to toluene. It smells and tastes like almond and is used as artificial almond flavoring. Real almond nuts also contain benzaldehyde. That's what makes them taste like almonds. If benzaldehyde can be produced from either a "natural" source (a nut), or an "artificial" source (chemicals in a lab), should benzaldehyde be classified as natural flavoring or artificial flavoring? That's a tricky question, isn't it? (Also, isn't it strange that the addition of the aldehyde group will change the way your liver treats the benzene ring? Almond flavoring is not toxic even though it has a benzene ring.)

Once again, we are going to leave further combinations of these two groups to your college professors. Moving right along...

What would happen if you mixed an alcohol with a carboxylic acid? Let's try it and see what happens. Let's mix glycerol with nitric acid:

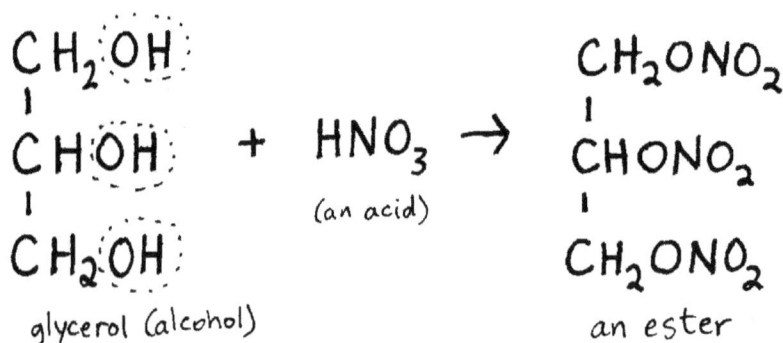

$$CH_2OH$$
$$CHOH \quad + \quad HNO_3 \quad \rightarrow$$
$$CH_2OH$$

(an acid)

glycerol (alcohol)

$$CH_2ONO_2$$
$$CHONO_2$$
$$CH_2ONO_2$$

an ester

I'm sure glad we don't have to memorize these!

We get... an ester? Yes, an ester called **nitroglycerin**. You may be familiar with the name of this chemical. You may know that it is used to make dynamite (TNT). What you may not know is that nitroglycerin is also a medicine used to treat cardiovascular disease.

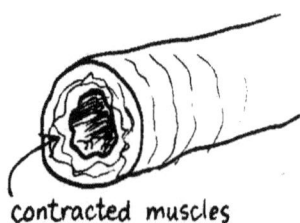

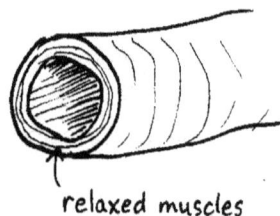

contracted muscles

relaxed muscles

Nitroglycerin causes the muscles that line the inside of blood vessels to relax. The arteries open up, and then more blood can get through.

An alcohol and a carboxylic acid can combine to form an ester. But can you go the other way? Can an ester be broken apart to make an alcohol and an acid? Yes, and in fact, people have been doing this for centuries, although they did not know the chemistry behind what they were doing. They were simply making soap. Before the age of high-tech detergents, soap was made by heating animal fat with lye. Lye was made from wood ashes.

A fat called glycerol tristearate can be broken down into stearic acid and glycerol:

This is not for the
faint of heart!
But you're tough!

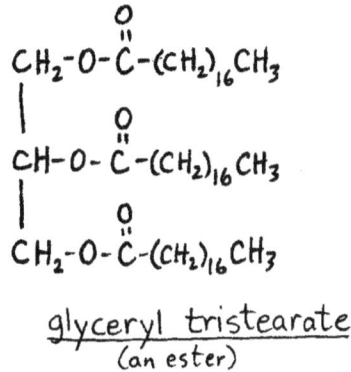

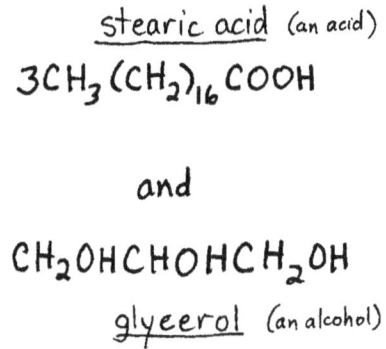

$$CH_2-O-\overset{O}{\overset{||}{C}}-(CH_2)_{16}CH_3$$
$$CH-O-\overset{O}{\overset{||}{C}}-(CH_2)_{16}CH_3$$
$$CH_2-O-\overset{O}{\overset{||}{C}}-(CH_2)_{16}CH_3$$

glyceryl tristearate
(an ester)

$\xrightarrow{\text{NaOH (lye)}}$

stearic acid (an acid)

$$3CH_3(CH_2)_{16}COOH$$

and

$$CH_2OHCHOHCH_2OH$$

glyeerol (an alcohol)

The stearic acid then reacts with the lye (sodium hydroxide) to form sodium stearate, a type of soap. The sodium stearate floats to the top of the cooking mixture and can be skimmed off.

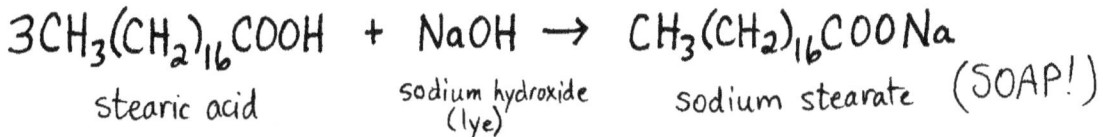

$$3CH_3(CH_2)_{16}COOH + NaOH \rightarrow CH_3(CH_2)_{16}COONa$$

stearic acid     sodium hydroxide     sodium stearate    (SOAP!)
(lye)

So that's how you make old-fashioned soap. But how does soap work? The secret of the soap molecule is that its ends are very different: one end loves oil (or greasy dirt), and the other end loves water.

We could call this the "lollipop" molecule!

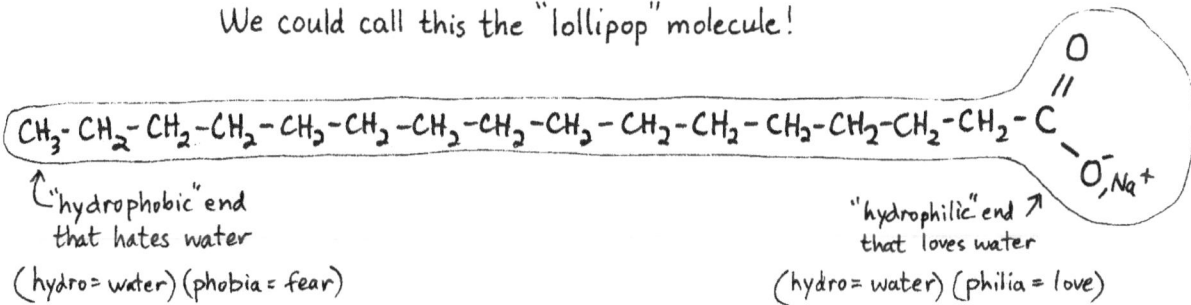

$$CH_3-CH_2-CH_2-CH_2-CH_3-CH_3-CH_2-CH_2-CH_3-CH_2-CH_2-CH_2-CH_2-CH_2-CH_2-\overset{O}{\overset{||}{C}}-O^-,Na^+$$

"hydrophobic" end
that hates water
(hydro= water) (phobia = fear)

"hydrophilic" end
that loves water
(hydro= water) (philia = love)

All dirt particles are surrounded by a thin layer of oil.  Since oil and water don't mix, this makes it impossible for the water to "pick up" dirt particles and carry them away.  Every time a water molecule comes close to a dirt molecule, they repel like magnets do when you try to put two north poles together.  The water can't even get near the dirt.

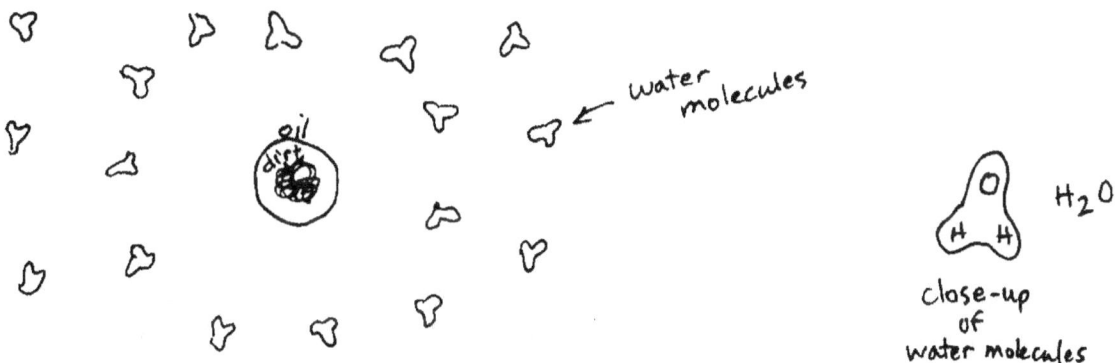

water molecules

$H_2O$

close-up
of
water molecules

Then along comes a soap molecule. The soap's oil-loving end grabs the dirt molecule and the soap's water-loving end grabs a few water molecules. The water molecules follow other water molecules rushing by, and the whole chain of water molecules goes down the drain.

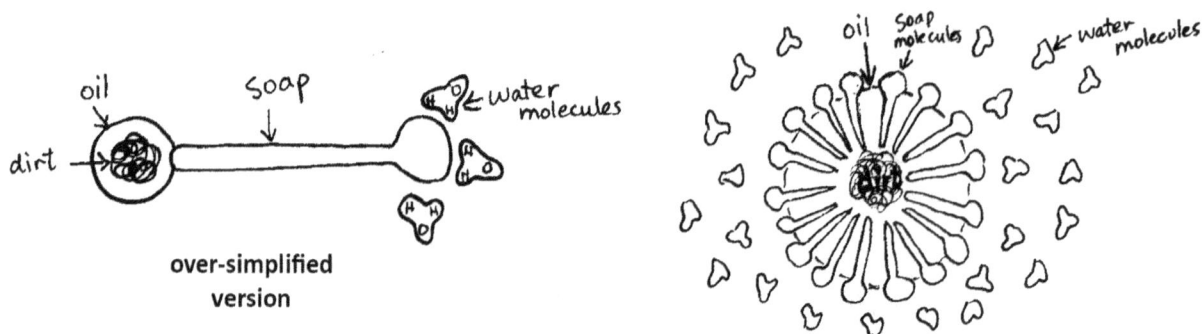

over-simplified
version

Soap molecules form a sphere called a "micelle."

Functional groups can often be spotted hanging off the ends of lots of large molecules. We are going to take a brief look at two types of large organic molecules found in animals: prostaglandins and pheromones.

**Prostaglandins** were first discovered in men and were thought to have come from the prostate gland (which is only found in males). Then…oops—they were found in females. So much for the name being appropriate. However, everyone had been using the name long enough that it was too late to change it. Prostaglandins control many body processes, such as blood pressure, body temperature, production of stomach acid, muscle contraction, pregnancy, the menstrual cycle, inflammation, and pain. Here is a diagram of a typical prostaglandin.

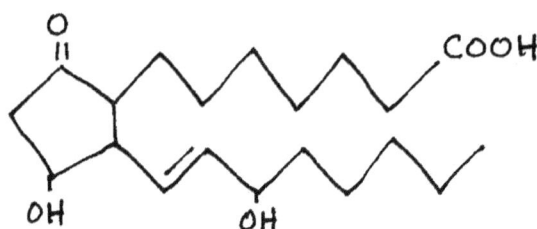

You will notice that there a lot of letters missing. Where are all the carbons and hydrogens? What do the lines mean? Each line is a bond, and at each point where two lines meet, there is a carbon atom. It's invisible, but all chemists know it is there.

Here is the same molecule with all the letters:

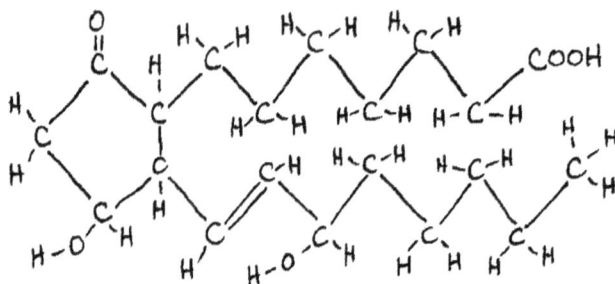

Yes, you can see why chemists prefer to leave out most of the C's and H's. No one has to remind them that carbons always make four bonds. They also know that oxygen (O) makes two bonds, but the O's are often shown, anyway.

Here are some other prostaglandin molecules, just so you can see what they look like:

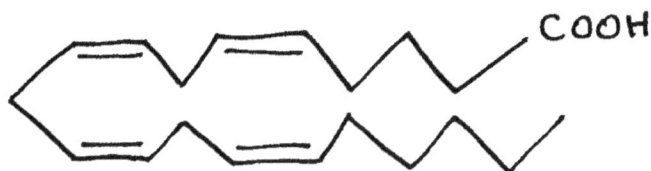

arachidonic acid

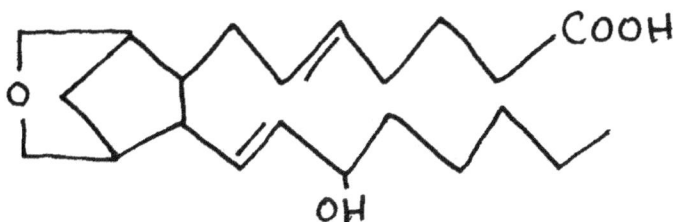

prostaglandin H2

Prostoglandins inspiring a work of abstract art:

The last type of molecule we are going to look at is a **pheromone**. Pheromones are the chemicals animals use to communicate. Animals "smell" pheromones and know what they mean. Here are some examples. Can you spot some functional groups? You might want to circle them with your pencil.

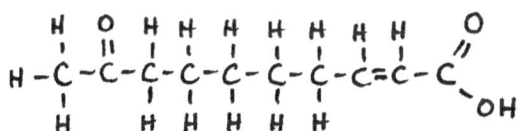

queen bee pheromone

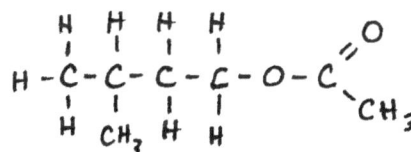

honey bee pheromone

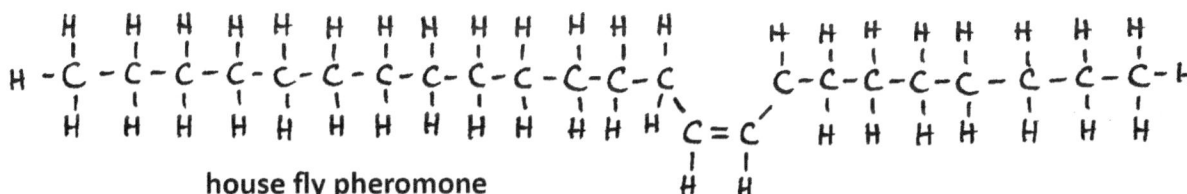

house fly pheromone

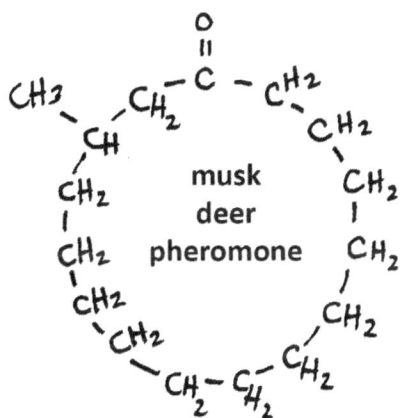

musk deer pheromone

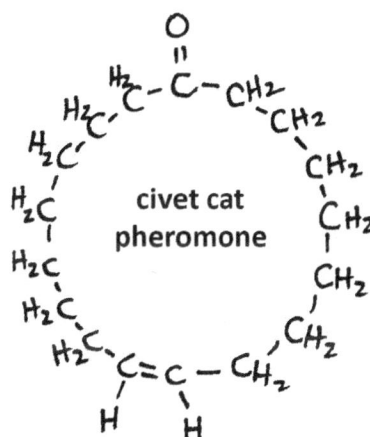

civet cat pheromone

Here are some examples of what pheromones can do:

- Ants secrete pheromones to warn of danger, to map a trail to a food source, to mark out boundaries, and to tell others what work needs to be done. Dead ants give off a pheromone that tells the other ants to carry it away. If you dab this pheromone on a live ant, its buddies will carry it off to the ant graveyard again and again, no matter how many times it returns! (Don't worry, the pheromone will wear off and the ant will be able to return to ant society.)
- Queen bees make pheromones to communicate with their drones.
- Some spiders make pheromones that smell like moth mating pheromones. A male moth thinks it smells a female, goes to the place where the smell is coming from, and gets stuck in the spider's web.
- Mammals have pheromones in their urine that lets them "mark" their territory with their scent. That's why you see them peeing on everything.
- Some plants make pheromones to attract a particular type of insect.
- Female rabbits release a pheromone that causes their babies to start nursing right away. The mother might have to flee at any moment, so the babies can't waste any time.
- Most animals make a series of pheromones that are used in the reproductive process. They communicate things like being in heat, being pregnant, etc.
- Male deer can smell female deer from miles away. It only takes a few pheromone molecules in the air for them to detect the scent!

Humans make pheromones, too, but we cannot smell them like animals can. They may still affect our brains, though, causing changes in our behavior.

---

What happens if we slightly alter the benzene ring itself? What if we removed one of the carbons and replaced it with a nitrogen?

This is called a **pyridine**. It is found in some B vitamin molecule, and in nicotine. Pure pyridine is a flammable, toxic liquid with a bitter taste and fishy smell. Its chemistry changes, though, if you add other atoms or molecules to it. If you attach it to a benzene ring you get quinoline, which is made into quinine, an anti-malaria drug.

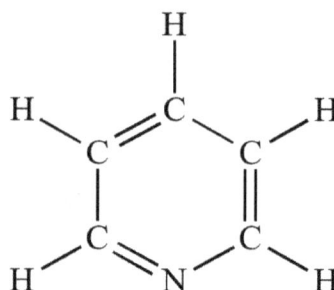

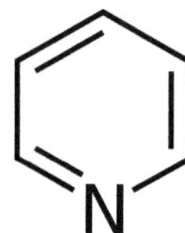

If you took away a carbon and made it a pentagon instead of a hexagon, you get something called a **pyrrole ring**. It is found lots of places: chlorophyll in plants, hemoglobin in blood, and B vitamins, just to name a few.

On the next page you will meet a molecule that has a benzene ring attached to a pyrrole ring. This makes a poisonous substance called "indole," that can be used in tiny doses as a medicine.

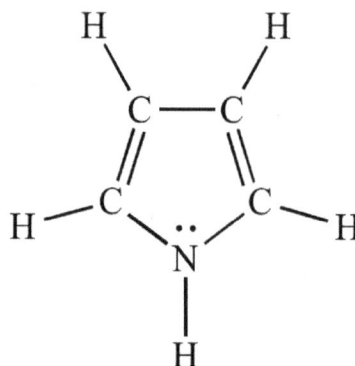

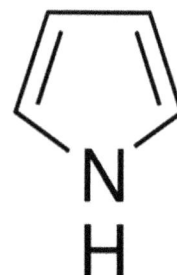

# THE INSPIRING STORY OF PERCY JULIAN

Percy Julian was born in Alabama, USA, around the year 1900. His grandparents had been slaves. Even though slavery had been abolished, blacks in the south still were not permitted to do many things, including attending good schools. Percy had to go to a high school that had very little money, and, therefore, had no science equipment. This was devastating for an intelligent young man whose heart's desire was to become a chemist.

Percy was a hard worker, though, and managed to get into DePauw University in Indiana. Even though he had to work full-time to put himself through college, he still managed to graduate at the top of his class. Because he was black, however, no one offered him any scholarships to go to graduate school. He had to take a teaching job and wait for an opportunity to come along.

When Harvard University sponsored a chemistry competition, Percy entered and won. He was then able to attend Harward, and he earned a Masters degree in chemistry in only one year. But still, no university would hire him.

Then one of his friends found someone to sponsor him to study in Vienna, Austria. He enjoyed working there at the University of Vienna, because no one cared about the color of his skin. While in Vienna, he heard about a chemical mystery no one could solve. There was an African bean, called the Calabar bean, that produced a substance doctors used to treat an eye disease called glaucoma. Scientists needed a way to produce this substance artificially, without having to rely on a supply of beans.

The Calabar bean had been used in Africa in witchcraft trials. They made suspected witches swallow it, thinking that if they died they were guilty and if they lived they were innocent. What was really going on was that the toxins inside the bean would shut down the person's nervous system, causing death. If the person was lucky enough to vomit up the bean, they would survive.

Percy returned to the United States to begin working on finding a way to make this substance, called physostigmine *(fizz-o-stig-mean)*. He knew that the empirical formula for physostigmine was: $C_{15}H_{21}N_3O_2$. He figured out that part of the structural formula must have two attached rings: a benzene ring and a pyrrole ring:

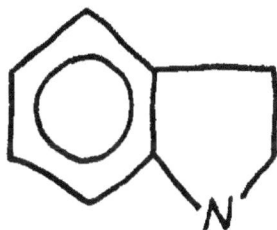

"indole"

This structure has a name of its own: ***indole***. In a diluted form it smells like orange blossoms. In a concentrated form it smells like raw sewage!

Percy could get indole from sources other than the Calabar beans, including soy beans, orange blossoms, and coal tar. He would use indole as the base for building an artificial physostigmine molecule. Next, he needed to find a way to attach the other necessary atoms to indole.

It took years of research, lots of money from a few generous donors, and help from a number of assistants before Percy finally arrived at the solution. He knew that another team of researchers in England was working on the same thing and had announced that they were very close to a solution. When Percy read their research report, though, he realized that they were on the wrong track and were not as close to a solution as they thought they were. Still, Percy needed to hurry so he could be first and could claim credit for the discovery.

At last he made a sample of what he believed was pure artificial physostigmine. To determine if he was right, he needed to compare it to natural physostigmine. If the artificial substance melted at exactly 139° F, just like the natural substance did, that would prove they were identical.

They watched the test tubes nervously, waiting to see if the two substances would melt at exactly the same instant, as soon as the temperature reached 139°. Yes! They did! Percy had made artificial physostigmine.

For this accomplishment, Percy won several prizes and honorary degrees. You'd think that now every university in the country would want him, right? No, they all turned him down, even his alma mater, DePauw University.

Percy took a job with Glidden Company in their soy bean research department. He discovered how to make artificial hormones and a drug called cortisone from soy beans. He also invented fire-fighting foam. He eventually started his own company, and ended up being a millionaire.

This story ends happily ever after, but don't forget that the reason it did was because of Percy's excellent character traits. When life was unfair to him, he did not whine and complain, but worked hard and proved himself.

Every one of us has times in our life when we feel disadvantaged in some way. Remember Percy's example, and "percy-vere"!

## Comprehension self-check

See if you can answer these questions.  If you can't, go back and look up the answer.

1) Name a molecule that has more than one carboxyl group on it: _____

2) What do acids put into a solution? Electrons, protons or neutrons?_____

3) What do you get when you put the carboxyl group on a benzene ring? _____

4) Name a product you use that might contain sodium benzoate: _____

5) What do you get when you put an aldehyde group on a benzene ring? _____

6) What does benzaldehyde taste like? _____

7) What functional group does nitroglycerin belong to?  _____

8) Name two ways nitroglycerin is used: _____, _____

9) What did people in past centuries use to make soap? _____ and _____

10) Where did people get the lye for soap? _____

11) Which one of these molecule might be called "soap"?

    a) sodium benzoate   b) sodium stearate   c) sodium hydroxide   d) sodium chloride

12) Is dirt normally attracted to water molecules? _____

13) Which end of the soap molecule is attracted to dirt?  hydrophobic or hydrophilic? (circle one)

14) Which of these functions is not substantially controlled by prostaglandins?

    a) body temperature   b) muscle contraction   c) inflammation   d) vision   e) pain

15) Give three examples of how animals use pheromones:

    a) _____

    b) _____

    c) _____

16) What do you get when you replace one of benzene's carbons with a nitrogen atom?

    a) benzoic acid   b) pyridine ring   c) pyrrole ring   d) indole

17) What do call a pentagon with only four carbon atoms plus a nitrogen atom?

    a) physostigmine   b) pyridine ring   c) pyrrole ring   d) indole

18) The substance physostigmine occurs naturally in : _____

19) Indole is made of these two rings: _____ and _____.

20)  The way Percy Julian knew that his substance was identical to natural physostigmine was by comparing their _____.


REVIEW QUESTIONS:

21) What does the prefix "but-" mean? _____

22) A hydrocarbon with at least one double bond is called an _____.

23) What do you call molecules that have the same number and type of atoms but differ in shape? _____

24) Diamond, graphite and coal are _____ of carbon.

25) How many electrons does a carbon atom want to have in its outer shell? _____

## Word puzzle (includes review from past chapters)

An aldehyde that tastes like almond: __ __ __ __ __ __ __ __ __ __ __ __
                           42        26      1  15   58

An explosive that can help the heart: __ __ __ __ __ __ __ __ __ __ __ __ __
                41   50 53 5      30    36      57

Physostigmine treats this eye disease: __ __ __ __ __ __ __ __
               7   45     38 39

This regulates body temperature: __ __ __ __ __ __ __ __ __ __ __ __
           34      10 61   14 4      28   6

$C_6H_{12}$ is this kind of formula: __ __ __ __ __ __ __ __ __
            19 51    37

Moths can smell these for miles: __ __ __ __ __ __ __ __ __
          16 22    20   25   59 49

Benzene with one carbon replaced with a nitrogen: __ __ __ __ __ __ __ __
                   46 18 44       33

The food preservative sodium benzoate is made from this acid: __ __ __ __ __ __ __
                29 27 60   24

Hydrophilic means loving: __ __ __ __ __
        3  9    13

The functional group OH makes something into an __ __ __ __ __ __ __
                  2  12 40

Benzyl acetate smells like: __ __ __ __ __ __ __
       47   17 55    52

Toluene is made by this tree: __ __ __ __
       11    48

Methyl benzoate smells like this fruit: __ __ __ __
        8

Cream of tartar (a baking ingredient) comes from this acid: __ __ __ __ __ __ __
        21     31

How many bonds can carbon make? __ __ __ __
       23 54

Compound W wart remover smells like this: __ __ __ __
       43 32

A benzene ring is made of carbon and __ __ __ __ __ __ __ __
       35     56

\*\*\*\*\*\*\*\*\*\*\*\*\*\*\*\*\*\*\*\*\*\*\*\*\*\*\*\*\*\*\*\*\*\*\*\*\*\*\*\*\*\*\*\*\*\*\*\*\*\*\*\*\*\*\*\*\*\*\*\*\*\*\*\*\*\*\*\*\*\*\*\*\*\*\*\*\*\*\*\*\*\*\*\*\*\*\*\*\*\*\*\*\*\*\*\*\*\*\*\*\*

__ __ __   __ __ __ __   __ __ __   __ __ __   __ __ __ __ __
1  2  3    4  5  6  7    8  9  10   11 12 13   14 15 16 17 18

__ __ __ __   __ __ __ __ __ __   __ __   __ __ __
19 20 21 22   23 24 25 26 27 28   29 30   31 32 33

__ __ __ __ __ __ __ __ __   __ __ __ __?
34 35 36 37 38 39 40 41 42   43 44 45 46

__ __ __ __   __ __ __ __ __ __ __ __ __-__ __ __ __!
47 48 49 50   51 32 52 53 54 55 56 57 58  55 59 60 61

# CHAPTER 6: PLASTICS

In this chapter, we will begin to combine what we learned about "anes" and "enes" with the things we learned about functional groups. We will be able to make some really cool items!

Do you remember good old ethylene? Well, if you play construction set with a whole bunch of ethylenes, and hook about a thousand of them together to make a long chain, you get a molecule called **polyethylene**. ("Poly" means "many.")

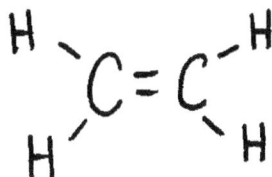

Remember ethylene?

Ethylene

## THREE WAYS OF SHOWING POLYETHYLENE:

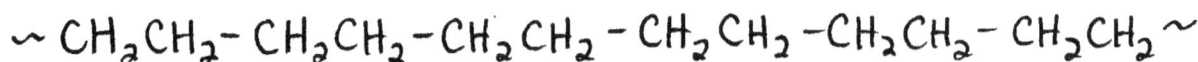

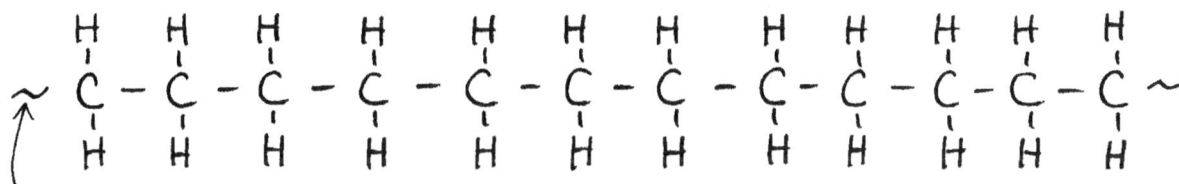

$$\sim CH_2CH_2 - CH_2CH_2 - CH_2CH_2 - CH_2CH_2 - CH_2CH_2 - CH_2CH_2 \sim$$

This is the organic chemistry symbol for "etcetera."

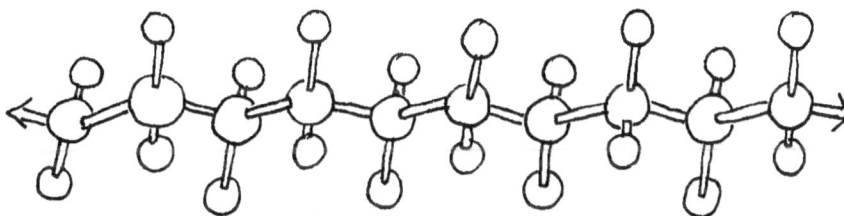

Don't forget — the molecule isn't flat. It has a zig-zag shape!

To make ethylenes form a long line, chemists use high temperatures, high pressure, and chemicals called **catalysts**. A catalyst is a chemical that helps a reaction to occur, or at least speeds it up. The catalyst is not destroyed in the process, so it can be used over and over again. Here is a silly way to remember what a catalyst does.

The same clergyman can join couples again and again.

The same catalyst can join molecules again and again.

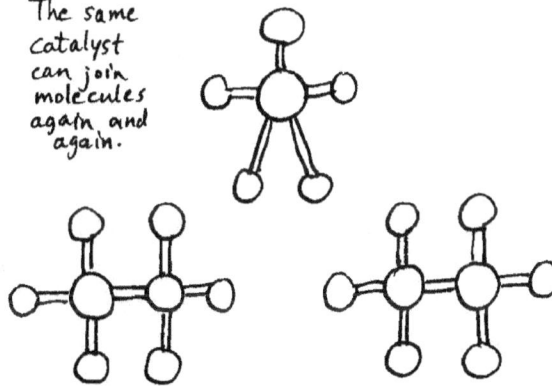

What can you do with polyethylene? Almost anything! You can make bags, bottles, jugs, cups, utensils, toys, sports equipment, tubes, insulators, photographic film (the stuff no one uses anymore because we've gone digital), beauty products, fabrics, and a lot more. Polyethylene is the most common type of plastic we use everyday.

Polyethylene was invented before the start of World War II and was immediately put to use by the Allied army for insulating cables. Polyethylene was flexible and tough and could withstand both high and low temperatures, making it perfect for protecting the wires that ran into important things like radar machines.

There are basically two types of polyethylene: high density and low density. **High density polyethylene (HDPE)** is made of long, linear chains that don't have many side branches. They can pack into an orderly crystalline structure, making high density polyethylene tough and strong, and therefore perfect for making things like bottle caps, toys, and milk jugs.

It didn't take a lot of talent to do that illustration!

**Low density polyethylene (LDPE),** on the other hand, has lots of side chains branching off the main chain. These branches prevent the molecules from packing close together.

Low density polyethylenes are more bendable and melt at lower temperatures. If you put a high density polyethylene object in boiling water, it will hold its shape. If you put a low density polyethylene object into boiling water, it will melt and become severely deformed. LDPE isn't inferior to HDPE, however. You really want bendable plastic for some things. Low density polyethylene is great for making squeeze bottles that will hold liquid products such as honey, mustard, shampoo, paints and lotions. It can also be rolled into very thin sheets and made into bread bags or produce bags. (Just to confuse use, HDPE can also be made into bags.)

Polyethylene plastic is a ***thermoplastic*** polymer, meaning that it can be melted and reshaped again and again. Not all plastics are like this. (***Thermosetting*** plastics have the opposite characteristic: they are "set" into a shape the first time, and that's it.)

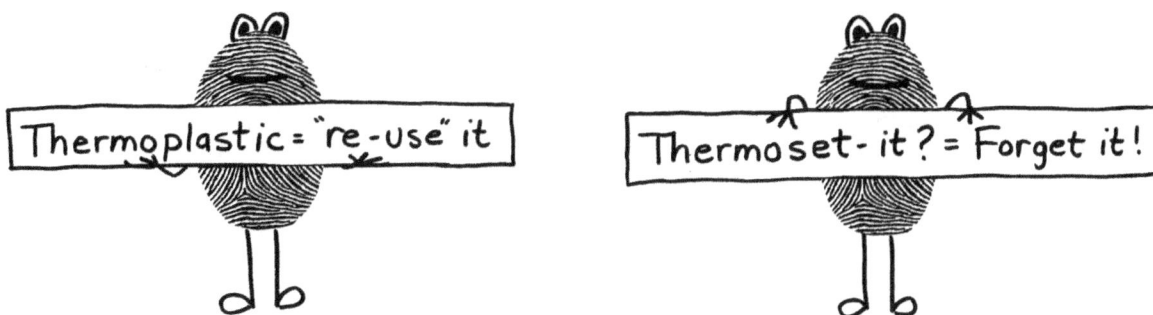

Because polyethylene can be re-shaped, it is easily recycled. For instance, old soda bottles and milk jugs can be turned into fabric (polyesters), park benches, play equipment, sports and camping gear, insulation for buildings, "bedding" for new highways, and much more. Plastics always have recycling numbers printed or embossed on them because the method of recycling will depend on the chemistry of the plastic. HDPE is number 2, and LDPE is number 4.

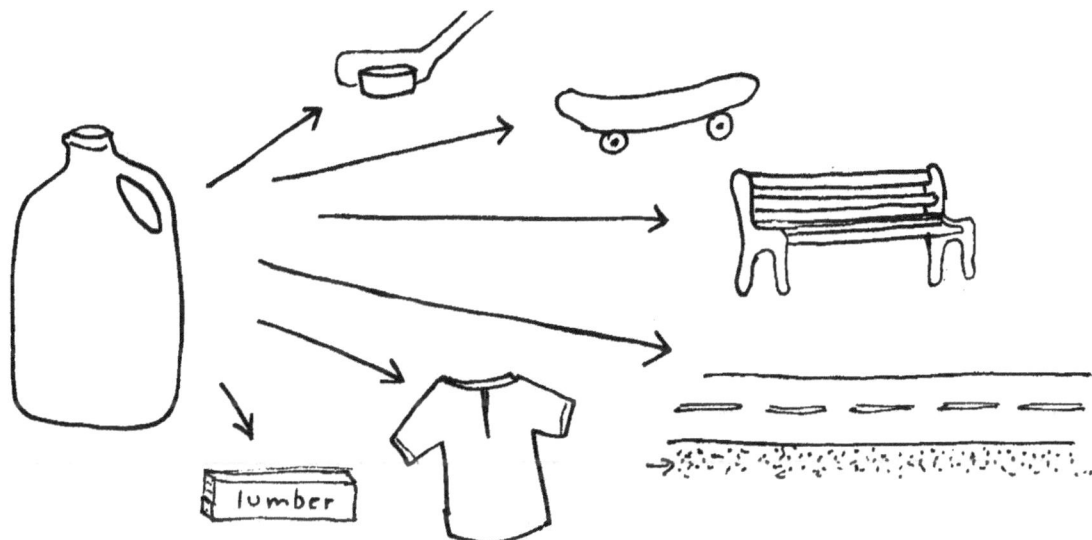

Polyethylene is just one of a whole category of molecules called ***polymers***. We already learned that "poly" means "many." "Mer" means a single molecular unit. ("Poly-olecule" or "polymoly" might have been a better name, but perhaps these sounded too cute-sy for scientists.) Polymers are long chains of any type of molecule. There are natural polymers as well as man-made polymers. Starches and proteins are natural polymers.

Another man-made polymer is ***polypropylene***. The basic unit of polypropylene is propylene (or "propene" if you use the official IUPAC word). "Prop" means three carbons, and "ene" means there is a double bond in there somewhere, so it would look like this:

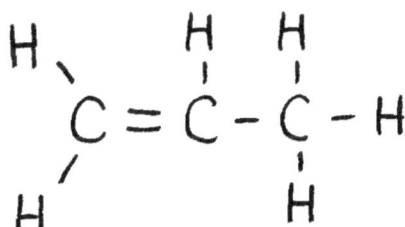

Now to make polypropylene into a polymer. We will open up the double bond and rearrange the hydrogens a bit. A propylene will attach to every other carbon atom, with hydrogens in between.

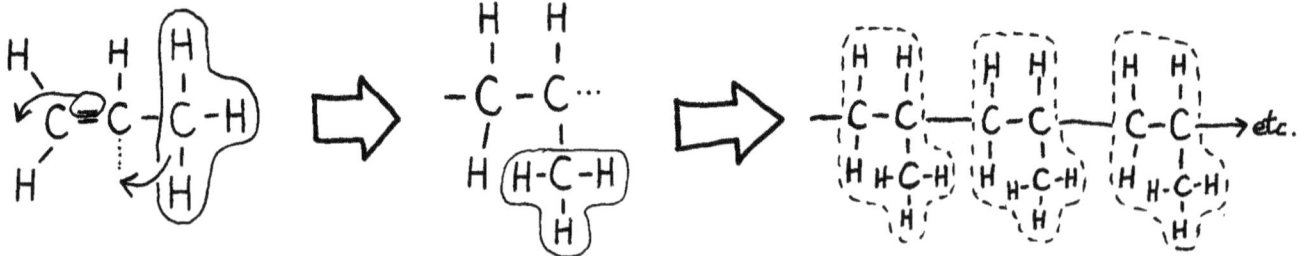

Propylene gets rearranged.

This is the "mer" (one unit of propylene).

Then join them to make polypropylene.

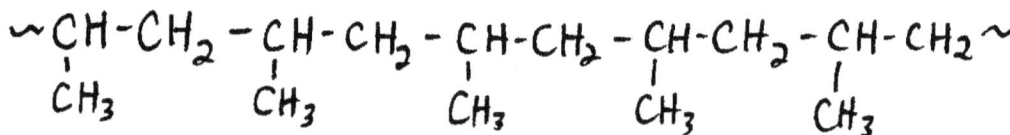

$$\sim CH-CH_2 - CH-CH_2 - CH-CH_2 - CH-CH_2 - CH-CH_2 \sim$$
$$\quad CH_3 \qquad\quad CH_3 \qquad\quad CH_3 \qquad\quad CH_3 \qquad\quad CH_3$$

*Speaking of hard-shell luggage, I need a vacation from this stuff!*

Polypropylene is a tough plastic suitable for molding into hard-shell luggage, battery cases, and appliances. In small pieces it is good for indoor-outdoor carpets.

Let's do a crazy experiment! Let's attach one of those nifty benzene rings onto a polyethylene chain. What will happen?

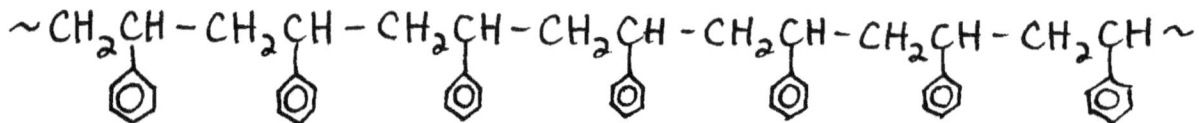

$$\sim CH_2CH - CH_2CH - CH_2CH - CH_2CH - CH_2CH - CH_2CH - CH_2CH \sim$$

*I think they look like icicles.*

*They remind me of wasp nests – or maybe bird houses.*

Yikes! We've made **polystyrene**! We've helped to pollute the world! Add some air bubbles and we've got Styrofoam™ to throw into landfills: disposable hot beverage cups, fast food plates, padding inside appliance and toy boxes, packing "peanuts," etc. Of course, Styrofoam is wonderful when you want things insulated and padded. That's why they make it. But it does create a long-term garbage problem.

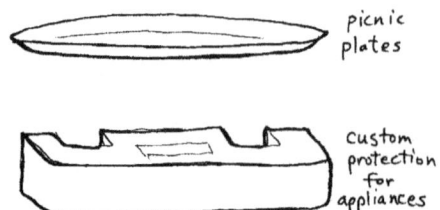

*Airhead!*

*Look who's talking, foam face!*

*picnic plates*

*custom protection for appliances*

What will happen if we add some chlorine molecules to polyethylene?

$$\sim CH_2CH - CH_2CH - CH_2CH - CH_2CH - CH_2CH - CH_2CH \sim$$
$$\quad | \qquad\qquad | \qquad\qquad | \qquad\qquad | \qquad\qquad | \qquad\qquad |$$
$$\quad Cl \qquad\quad Cl \qquad\quad Cl \qquad\quad Cl \qquad\quad Cl \qquad\quad Cl$$

Chemists call this **polyvinyl chloride**. It's perfect for many things. You can make it into artificial leather. You can make it into pipes that will never rust. You can make it into unbreakable clear bottles, floor tiles, shower curtains, plastic wrap, toys, hardware parts, and much more. It's amazingly useful.

This is fun. Let's mess up polyethylene some more! Let's add fluorine instead of chlorine. We'll adapt the ethylene molecule first, by putting flourines on instead of hydrogens. That'll shake things up! What will happen?

We've made... Teflon™. Teflon is used for many things, but the one which you are probably most familiar with is the coating on non-stick frying pans.

Here's another idea. Can we replace three hydrogens with one nitrogen? Nitrogen can make three bonds. Three hydrogens should be equal to one nitrogen. Let's pop off three H's and stick on an N and see what happens.

Wow! We've created something you artsy-craftsy people will like. This substance is called "polyacrylonitrile." Don't bother trying to remember it, let alone pronounce it, but it makes great yarn, and is a key ingredient in many paints.

# SAVE THE ELEPHANTS!

Once upon a time, people used ivory to make things like piano keys and billiard balls. Since ivory comes from elephant tusks, and elephants do not go around giving away their tusks, the use of ivory necessarily involves killing elephants. This is not an ideal situation, needless to say.

You're not taking MY tusks, pal!

So a contest was held to see if anyone could come up with an ivory substitute. An American inventor named Wesley Hyatt found a way to soften cellulose nitrate by treating it with ethyl alcohol and camphor. (Cellulose is the "stuff" that plants are made of. When it is treated with nitric acid, it forms cellulose nitrate.) This new material, celluloid, could be shaped into smooth, hard billiard balls. This was also wonderful for the billiard table industry, because since these new balls were much less expensive than ivory ones, an average-income family could now afford a billiard table.

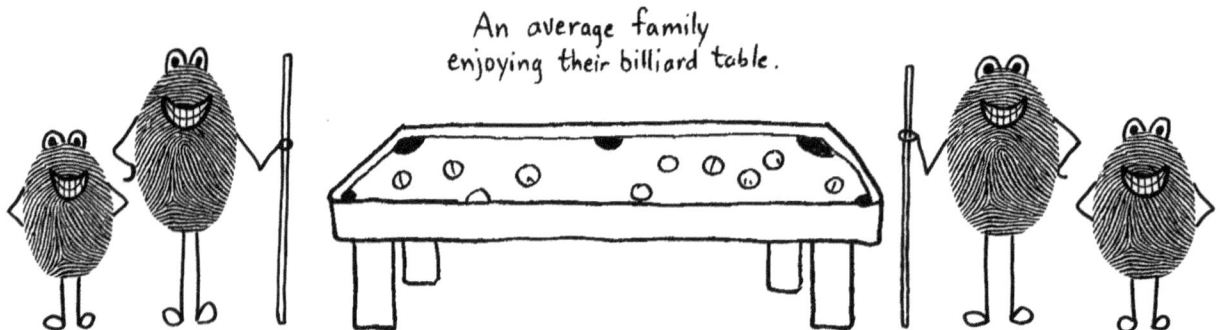

An average family enjoying their billiard table.

Celluloid was also used in the brand new film industry, just getting started in the early 20th century. The movie industry became known as the "celluloid industry."

Celluloid had the unfortunate characteristic of being highly flammable, however. Another use for it was smokeless gunpowder. Celluloid billiard balls could explode if you hit them hard enough. Celluloid film could also ignite spontaneously, and cause fires in theaters. As soon as a safer substance became available, celluloid was abandoned. Today, cellulose acetate is used instead of cellulose nitrate.

Ooops...

# SAVE THE LAC BUGS!

Once upon a time, people used a tiny bug called a lac to make "shellac," a substance that was painted onto surfaces made of wood to give them a shiny finish. As you can imagine, it took an awful lot of lac bugs to make enough shellac to varnish even a small picture frame.

Then along came Leo Baekeland. He came to the United States from his homeland of Belgium. After achieving great success in the photographic film industry, he used his riches to set up his own research laboratory. Leo put his lab to work trying to find a substitute for shellac.

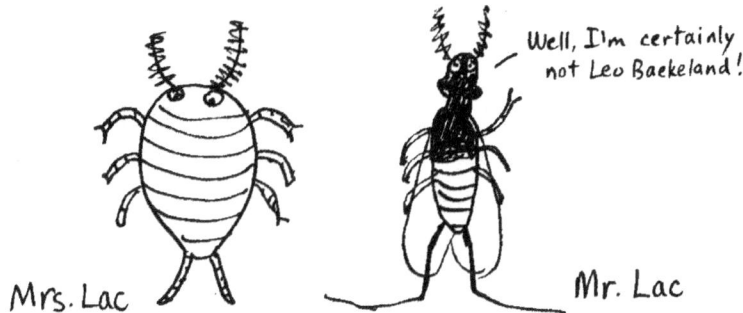

Well, I'm certainly not Leo Baekeland!

Mrs. Lac          Mr. Lac

One of the experiments he tried was mixing carbolic acid with formaldehyde. Other chemists had warned him about this experiment, but he did it anyway. When the mixture cooled, it turned into a hard solid that simply would not come out of the test tube. No matter what anyone tried, the substance could not be dissolved. Lots and lots of glassware had to be pitched at Leo's lab!

One day, Leo got to thinking that a substance that could not be dissolved by anything might actually be useful. He came up with a way to shape it into practical things like bowls. A bowl that would hold harsh chemicals and not dissolve would be helpful in chemistry labs. He knew he was on to something, so he decided to give his new substance a name: Bakelite. This was the world's first plastic.

Ta-da!

Bakelite was soon used for many purposes. One of the more well-known uses was for telephones. (You know, those black phones you see sitting on desks in old movies.) Bakelite plastic is still used today for things like light switches and pot handles, though the term "Bakelite" is seldom used.

## Comprehension self-check

You know what to do by now, right?

1) What does "poly" mean? _____     What does "mer" mean? _____
2) A chemical that helps a reaction occur, without itself being consumed is a _____.
3) How many carbon atoms are in a single unit of ethylene? _____
4) The two kinds of polyethylene are _____ _____ and _____ _____.
5) Which has branched molecules, HDPE or LDPE? _____
6) Which one will melt more easily, HDPE or LDPE? _____
7) Would a hard plastic bucket be more likely to be made of HDPE or LDPE? _____
8) Would a soft squeeze bottle be more likely to be made of HDPE or LDPE? _____
9) Is polyethylene a thermoplastic or a thermosetting plastic? _____
10) What recycling number is used for HDPE? _____
11) What is the single unit for polypropylene?
   a) ethylene     b) propane     c) propene     d) isopropol
12) Which one of these is most likely to be made of polypropylene?
   a) suitcases     b) bread bags     c) squeeze bottles     d) plastic shopping bags
13) What do you need to add to polyethylene to transform it into polystyrene?
   a) aldehyde group     b) benzene ring     c) alcohol group     d) carboxyl group
14) What is a brand name we use for a common type of polystyrene? S_____
15) What does PVC stand for? _____
16) A long chain of carbon atoms with fluorines attached is called T_____, a substance that is famous for its non-_____ surface.
17) Natural ivory is taken from the _____ of _____.
18) Why is celluloid no longer used? _____
19) Did Baekeland ever find a way to manufacture artificial shellac? _____
20) Which one of these would NOT be a use for Baekeland plastic?
   a) desk phone     b) plastic bag     c) pot handle     d) light switch


## Sing the Plastics Song!

The audio track is here:  **www.ellenjmchenry.com/audio-tracks-for-carbon-chemistry**

# The Plastic Song

(to the tune of "Big Rock Candy Mountain")

I'm proud of my collection, it's not toy cars or dolls;
I don't like coins or bottle caps, or stamps or baseball cards.
I'm proud to be the owner of many fine works of art;
      they are very lightweight,
      when the fall they don't break,
      can be made in any shape,
      come in clear and opaque,
and I dearly love my plastic!

I fell in love with plastic when I was just a kid,
I filled my shelves and closets with bottles, bags and lids.
I filled my head with knowledge and learned how plastic's made;
      it comes from oil
      then the oil they do boil
      making liquid and gas
      which then do pass
to the factories that make plastic.

I know that plastic's made of polymers so miniscule.
The poly part means many, the mer's a molecule.
The mers and made of atoms, with carbon at the core;
      a polymer's a chain,
      a very long chain,
      a molecular train
      that can take a lot of strain,
and polymers make up plastics.

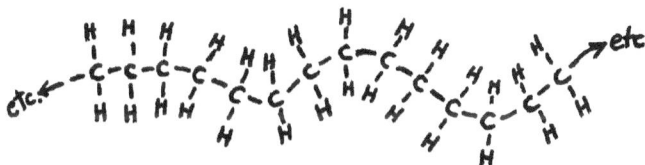

I've noticed little numbers on my plastic works of art;
I know they're for recycling (if I have a change of heart!)
The names that match these numbers-- they mostly end in -ene:
      like polyethylene
      and polypropylene
      and polystyrene
      but the one with chlorine
is polyvinyl chloride.

      So come to my home
      where you can roam
      in my Styrofoam
      and I'll let you take home
a precious piece of plastic!

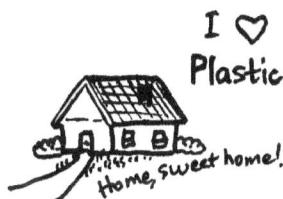

You can access the audio files here: **www.ellenjmchenry.com/audio-tracks-for-carbon-chemistry**

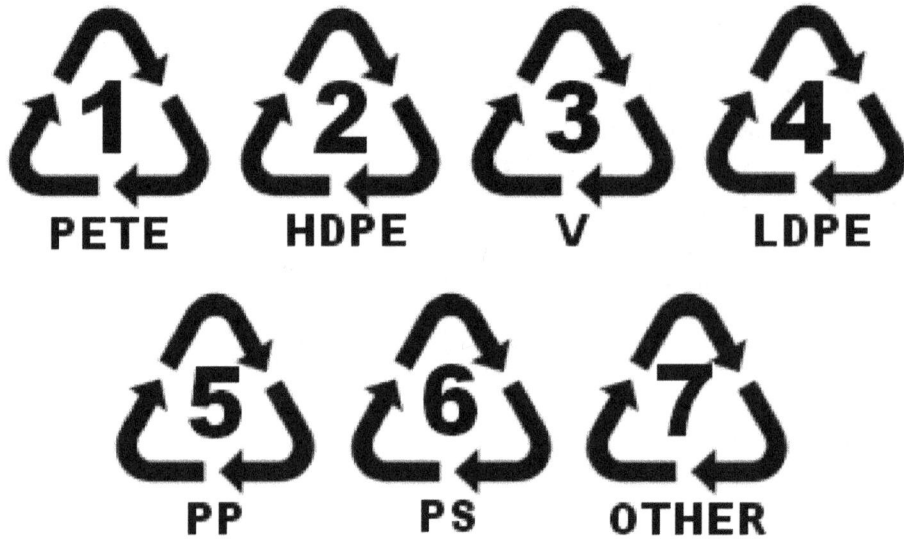

PETE = Polyethylene Terephthalate
HDPE = High-Density Polyethylene
PVC = Polyvinyl Chloride
LDPE = Low-Density Polyethylene
PP = Polypropylene
PS = Polystyrene

## Review

### Section 1: "Who Am I"?

| A | B | C | D | E | F | G | H |
|---|---|---|---|---|---|---|---|
| crude Oil | methane | chloroform | carbon tetrachloride | isomer | CFC's | acetic acid | methanol |

____ 1) Burning me puts nothing but carbon dioxide and water into the air, so they say I burn "cleanly."

____ 2) Scientists suspect that I might be damaging the Earth's ozone layer.

____ 3) I am a heavy liquid that is used in fire extinguishers and in dry cleaning of clothes.

____ 4) My name means "same parts." I have many twins who are identical to me and yet have different shapes.

____ 5) My nickname is "black gold."

____ 6) I can put people to sleep.

____ 7) I would be methane, except that one of the hydrogens has been replaced with an OH functional group.

____ 8) I am the active ingredient in vinegar.

### Section 2: Matching

____ 9) propane

____ 10) butene

____ 11) butyne

____ 12) ethanol

____ 13) naphthalene

____ 14) soap molecule

____ 15) acetic acid

____ 16) prostaglandin

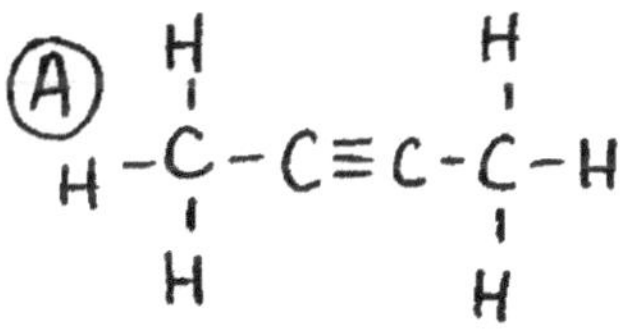

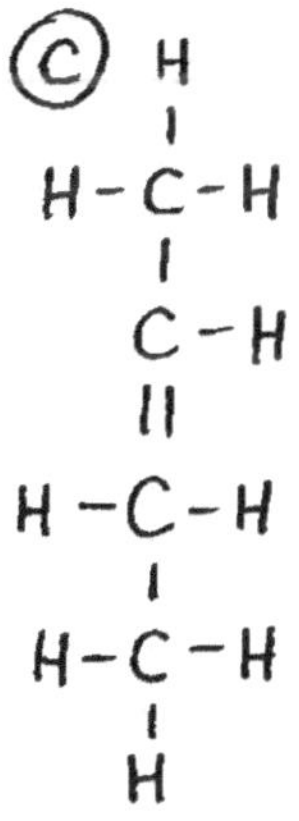

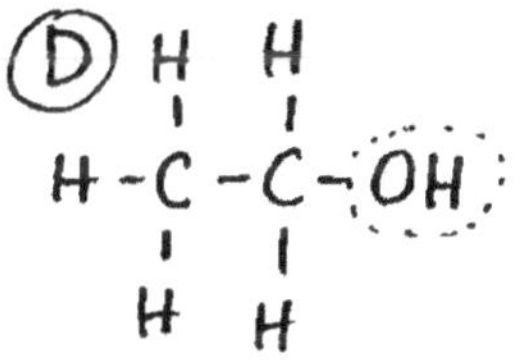

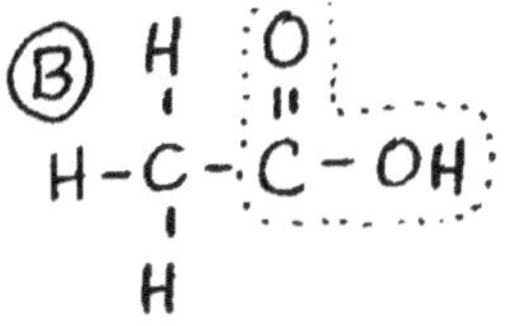

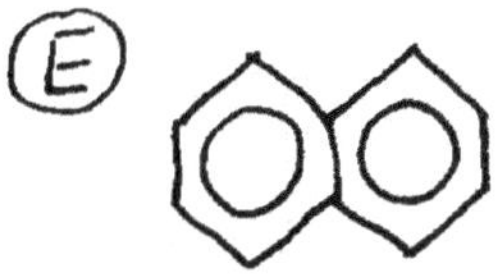

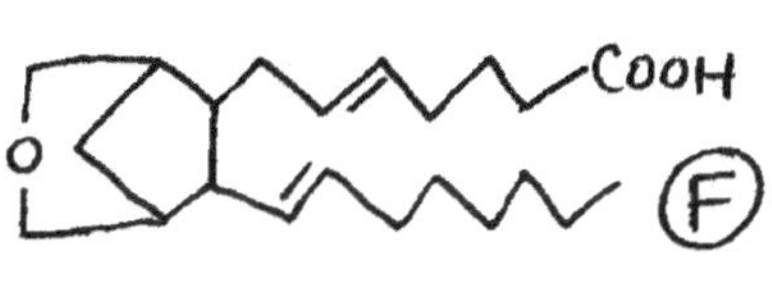

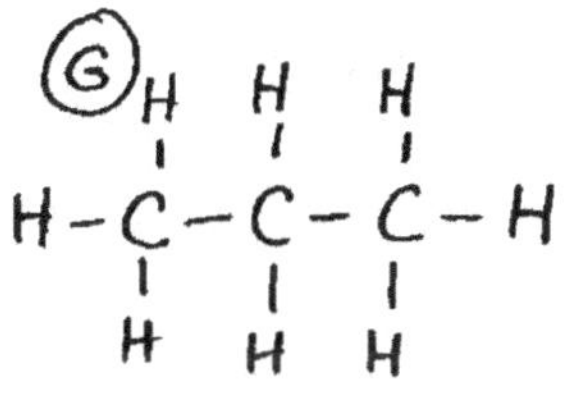

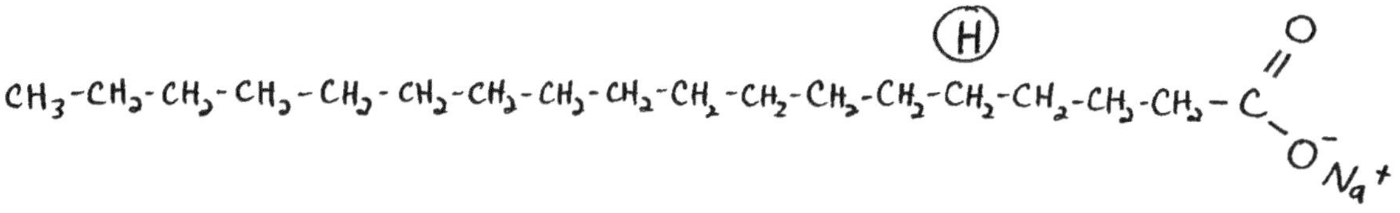

## Section 3: You respond

17) What does hydrophobic mean? _____

Name two ways animals use pheromones:

18) _____

19) _____

20) A carboxylic acid and an alcohol combine to form an _____.

21) Can you put more than one functional group on a molecule? _____

22) What does methane smell like? _____

23) Aromatic hydrocarbons are molecules that contain at least one _____ _____.

24) Formic acid is made by this member of the animal kingdom: _____

25) Formalin is used to _____.

## Section 4: True or False

___ 26) Esters are toxic.

___ 27) Animals can smell pheromones.

___ 28) Nitroglycerin is explosive.

___ 29) Nitroglycerin is used in heart medication.

___ 30) Lye is used to make soap.

___ 31) Ethers smell good.

___ 32) Ethers are used as artificial food flavorings.

___ 33) Ethanol is the type of alcohol found in beer.

___ 34) Acetone is toxic.

___ 35) Naphthalene is toxic to moths.

___ 36) Ethylene gas is toxic.

___ 37) Alkanes contain only single bonds

___ 38) Soap contains fat.

## Section 5: Draw

39-40) Draw a benzene ring without using any letters.   (2 pts.)

# CHAPTER 7: RUBBER AND SILICONES

Polymers occur everywhere in nature. Remember, the word "polymer" doesn't mean manmade, it only means a very long chain. The natural world is full of polymers. The first one we are going to look at is rubber.

Natural rubber was discovered in Central and South America hundreds of years ago by the native peoples who lived there. Certain trees oozed a milky substance that we now called *latex*. They used latex for making flexible containers, shoes, and balls, and also used it for waterproofing. When the Europeans came to the Americas, they were fascinated with it and sent samples back home to Europe. One of these samples came into the hands of a scientist named Joseph Priestly (who is famous for discovering oxygen, not for working with rubber). He found that the strange ball could rub off pencil marks, so he began calling it a *rubber*. More and more people acquired "rubbers" to erase their mistakes, and the name stuck even though many other uses were found for this substance.

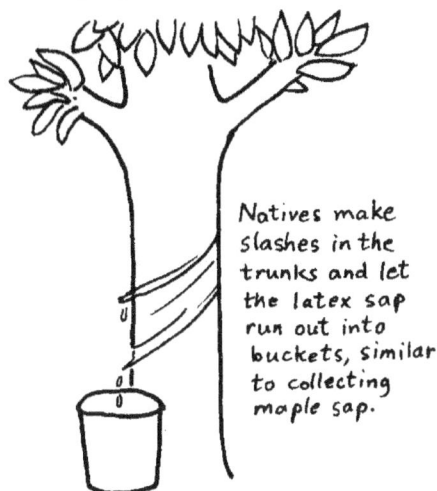

Natives make slashes in the trunks and let the latex sap run out into buckets, similar to collecting maple sap.

The world's first rubber processing factory was set up in Paris in 1803. The factory didn't actually make rubber from scratch, of course; it just processed the latex imported from Central and South America. The "raw" latex rubber had to be mixed and chopped to make the polymer chains a bit shorter. In 1823 Charles Macintosh began using latex rubber to make waterproof fabrics. Some people still use the word "macintosh" when referring to a raincoat.

In 1823, Charles Goodyear accidentally discovered a way to improve rubber. Up until now, rubber had the unfortunate characteristic of adapting to the climate: it got gooey in the heat and brittle in the cold. Goodyear accidentally dropped hot sulfur onto the rubber and found, to his surprise, that this cured the gooey/brittle problem. The addition of sulfur made the rubber resistant to the effects of both heat and cold. He called his accidental invention "vulcanization." (Vulcan was the Roman god associated with volcanoes, which emit a lot of hot sulfur.) Vulcanized rubber opened up whole new industries, such as the manufacturing of tires.

Poor Goodyear died in poverty. He never got to see tires with his name on them!

I like this page— not a chemical formula in sight anywhere!

The demand for latex rubber benefited countries like Brazil, where the climate is ideal for growing rubber trees. In fact, Brazil decided that it would not allow the latex trees to be taken out of the country to be grown elsewhere. Despite this law, seeds from these trees were smuggled to England, and from there were taken to Sri Lanka and Indonesia, where latex trees farms were started. In time, these tree farms produced so much latex that they became the world leaders in latex production, just as Brazil had feared.

To understand Charles Goodyear's accidental invention (vulcanization), we need to take a close-up look at the rubber molecule. Since it is a polymer, it must have individual "mer" units. The "mer" in rubber is called **isoprene**.

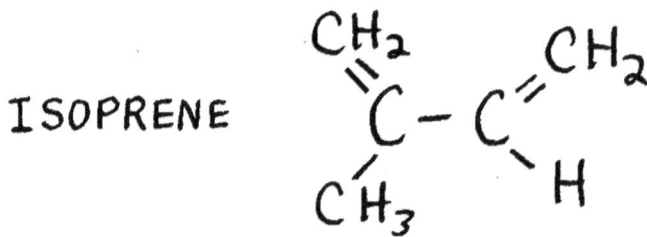

ISOPRENE

$$CH_2 = \underset{CH_3}{C} - \underset{H}{C} = CH_2$$

It's a litte more complicated than ethylene, but just a little. To make rubber, we need to attach a bunch of these isoprenes together:

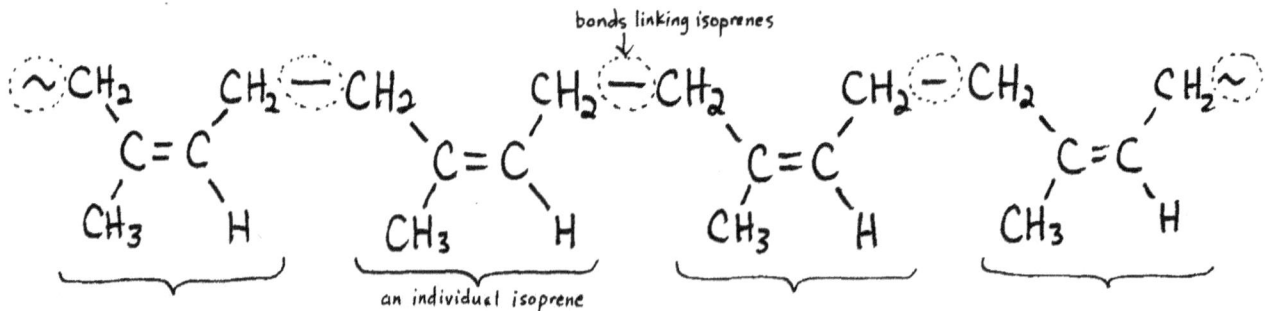

bonds linking isoprenes

an individual isoprene

Now we need to "zoom out" a little, so that we can't see the letters any more. We'll just draw rubber polymers with wiggly lines, like this:

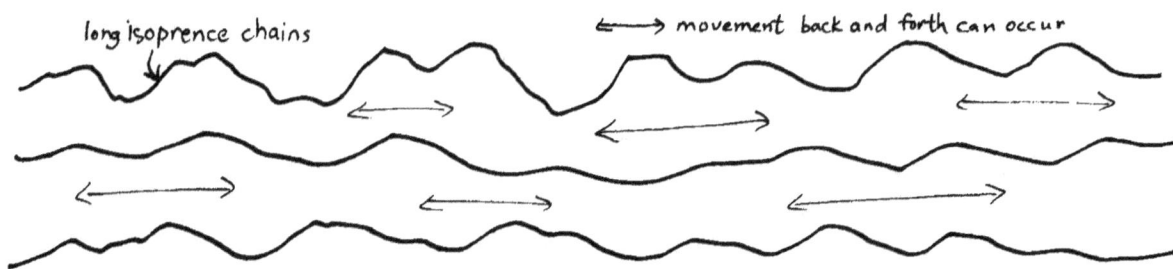

long isoprence chains

⟵⟶ movement back and forth can occur

The chains are not attached to each other and can slip back and forth. The slipping around of the chains is what causes natural rubber to be too soft. What happens during vulcanization is that the sulfur atoms connect the chains together at various points. Scientists call this **cross-linking**. The cross links stop the polymers from slipping. They also cause the rubber to snap back to its original shape after being stretched. Materials that stretch and snap back are called **elastomers**.

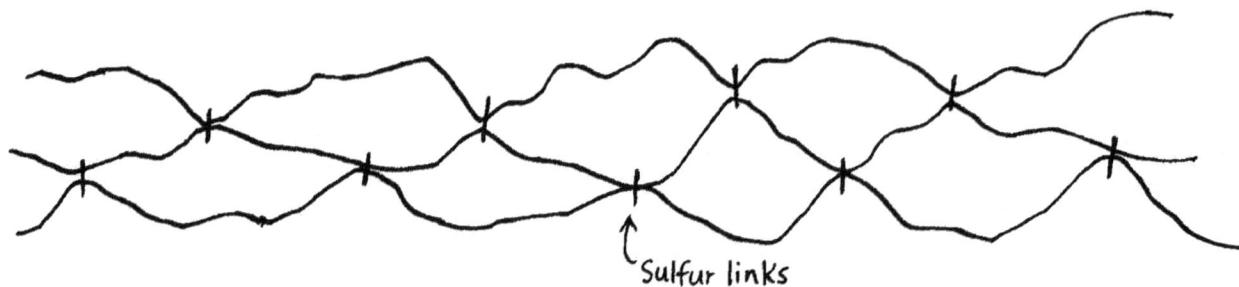

Sulfur links

During World War II, Japan blocked the Allied countries from receiving shipments of latex rubber from Indonesia. (No tires, no military vehicles!) America responded by trying to find a way to make synthetic (manmade) rubber. This search led to the discovery of other polymers such as Neoprene. After much experimenting, scientists finally found a way to make polyisoprene (artificial rubber) from petroleum. Today, 60% of all rubber comes from petroleum refineries, not latex trees.

We all know that rubber bands stretch; they are elastomers. Many fabrics stretch. Balloons stretch. But there are other things in our everyday life that contain elastomers, and we don't even know it. One surprising place you meet elastomers is in paint. Water-based latex paints contain rubber elastomers that make the paint flexible after it is dry. Chemists are always trying to improve the durability of the elastomers in paint so that we don't have to re-paint things so frequently.

Another type of natural latex rubber is **chicle**. *(CHICK-lay or CHICK-ul)* The chemical formula for chicle is very similar to rubber. Chicle comes from the sap of the sapodilla tree in Central and South America. The natives there liked to use chicle as chewing gum. In the 1800's, Americans began to investigate this natural rubber to see if it might be good for something. Nope. It was terrible as rain gear and even worse for tires. One day an inventor popped a piece in this mouth, and... thus began America's obsession with chewing gum. Food scientists discovered ways to add flavorings and colorings to make chewing gum even more enjoyable. Chewing gum was sold in vending machines in the New York subway as early as 1888.

Chiclets™ gum is named after chicle.

Bubble gum was not invented until 1906, by a man named Frank Fleer. His invention came on the market in 1926 with the name "Dubble Bubble." Bubble gum is not just chicle. It is a synthetic polymer similar to chicle, but with much better elasticity. Plain old chewing gum doesn't let you blow huge bubbles like bubble gum does.

Humans have always been chewing "gum." The ancient Greeks chewed mastiche *(mas-teek-uh)*, from the mastic tree. Ancient Mayans chewed chicle. Native Americans in North America chewed spruce sap, and passed the habit on to the settlers. The settlers improved the sap by adding beeswax. But nothing really compares with modern chewing gum that has textures and flavors perfected by food chemists.

Even though we are studying carbon chemistry, let's look at one polymer that is not based on carbon. It just seems to fit in so nicely that it is a shame not to mention it. This polymer is based on a chain of silicon and oxygen atoms, instead of carbon.

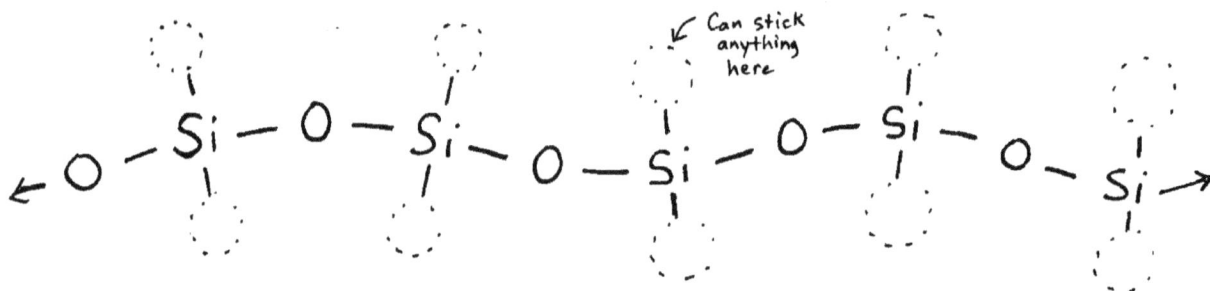

The reason silicon can replace carbon as the core of a polymer is because it makes four bonds, just like carbon does. (Look at a Periodic Table and you will see that silicon is right under carbon. All the elements in that column have the same ability to make four bonds.)

Silicones can be oily, rubbery, or solid, depending on how long the polymer chains are. Some silicon polymers are excellent for waterproofing. Raincoats and umbrellas are often coated with silicon polymers. Silicon oils are used as lubricants, and silicon-based fluids are used in hydraulic pumps. Polishes for cars and shoes often have a silicon base. The most famous silicon polymer is probably "Silly Putty."

You will remember that during World War II scientists began searching for a way to make synthetic rubber. One of these scientists was James Wright, who worked for General Electric in Connecticut. He tried mixing boric acid with silicone oil. It didn't make rubber, but it did make a most intriguing substance. It was soft and stretchy but not sticky. What could it be used for? He couldn't think of a use for it, so he sent samples to other scientists and asked if they could find something it was good for. After several years, not a single scientist had come up with a use for the stuff! Then a creative toy manufacturer got a sample of it and decided it was just fine the way it was. Pop it in a plastic egg and sell it for a few dollars. And thus, Silly Putty™ was born. It still doesn't have a practical use, but it's very fun to play with.

300 million eggs of Silly Putty have been sold since it was invented! Its debut in 1950 was in the spring, just before Easter, so they put the putty into plastic eggs. It has been in eggs ever since.

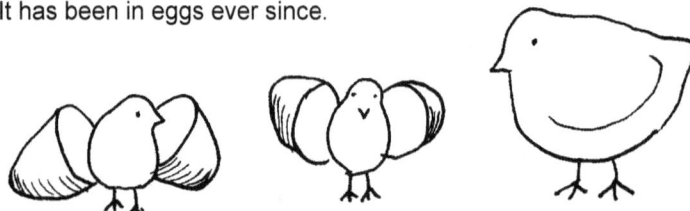

Silly Putty™ is actually a very *viscous* (thick) liquid. A substance with low viscosity is very runny, like milk or water. A substance with high viscosity is thick, like syrup. Silly Putty™ is even thicker than syrup. It is so viscous that it will take hours, even days, to drip off a table. Glass is also classified as a viscous liquid, taking centuries to drip. Old glass windows are thicker at the bottom.

**Don't forget that silicon polymers really don't belong in this book. Their chain is based on silicon, not carbon.**

## Comprehension self-check

1) Where were latex rubber trees first discovered? _____

2) Which one of these did natives NOT do with their natural latex?

   a) make balls    b) make flexible containers    c) make shoes    d) make raincoats

3) What do we call the objects that were originally called "rubbers"?

   a) galloshes    b) erasers    c) boots    d) chewing gum    e) bouncy balls

4) Who was the first person to use rubber raincoats? _____

5) Who discovered how to "vulcanize" rubber? _____

6) What element is used to vulcanize rubber? _____

7) What does vulcanization do to the quality of the rubber?

   a) makes it softer            b) makes it harder          c) makes into tires

   d) makes it resistant to heat     e) makes it resist changes due to temperature

8) Rubber trees were smuggled out of Brazil and planted in: _____

9) Why were rubber trees not planted in England?  (The text doesn't tell you this, just use common sense about plants and geography.) _____

10) What is the "mer" in rubber polymers?

   a) isoprene    b) isopropanol    c) ethylene    d) propene    e) silicon

11) Vulcanization causes the polymers in rubber to:

   a) become slippery         b) cross link        c) branch        d) become straight

12) Polymers that stretch and snap back are called _____

13) The word "synthetic" means:

   a) made from chemicals    b) made in a lab    c) manmade    d) combined

14) How many bonds can silicon form? _____

Name three things that silicon polymers are used for (other than Silly Putty™).

15) _____

16) _____

17) _____


18) Put these substances in order, from low to high viscosity (in your opinion):

ketchup, vegetable oil, cold syrup, water, Silly Putty™, shampoo, petroleum jelly (Vaseline)

_____

19) Can emperature affect viscosity?  (Think about warm maple syrup.) _____

20) Does Silly Putty™ contain carbon? _____

# "1-800-4REVIEW"

All the words in this search have to do with rubber. In fact, they are all written in this chapter somewhere. The word bank is in code, but it isn't a straightforward "number-to-letter" correspondence—that would be too easy. Instead, we've invented a code that will require you to do some critical thinking. The numbers correspond to letters on a phone pad. So the number 2 could be either A, B, or C.

Good luck! You'll need it! (Oh, all right. We'll give you one letter in each word.) Write the word above the numbers when you figure it out.

| | | |
|---|---|---|
| 35278663R | 765963R | 528E9 |
| R82237 | 244C53 | 486 |
| 4767736E | 847C687 | 7U5387 |

| | | | |
|---|---|---|---|
| 2R2945 | S27634552 | C76775465 | P7437859 |
| 466D9327 | 62246T674 | 8U5226492846N | 4N3663742 |

```
Y U J F U T G K W Y R I P C F V C B H D G U M T R X W C
L Q E V Y U Y G Z B P H I N D O N E S I A A M O R I S Y
Q J L D H D M W A L K U H Q E Y Z Q P O L Y M E R G W Y
L X A V N P Y N J U M O L I Y F C V H R I S O P R E N E
N K S Y W T J L K H W Y A G Z E D N V U E Z U Y D E E C
U F T M C K R K S O L L T S T S M F I K Q H G E U P P O
J W O Y M E E A J J K J E M R N B M S Y I U X V H H O R
R U M N T Y X T R L R W X M G F Z C C G H A R R U F E H
X F E K F O H K X Q V R L A I I S V O U H V U D Q B V J
G L R P R I E S T L Y E Z C W R L K U C T H B T Y H L N
V U L C A N I Z A T I O N I F H M P S O C R G Y Y F R P
F Z G A E J M C H I C L E N W Q B G C O R M K P K R F S
L Y O Q R H H G E A V S X T Q P Q P K Y O Y B G E U I Q
H V O S A P O D I L L A E O A Y X H B W S I M A L B C W
E L D A R H V G I Q X S S S H X E N N G S A H N W B W D
F A Y V H T U S J R U T I H T Y E J Y U L R F P P E H R
V H E I K H E Y V E B R A Z I L D O P G I B S W M R H D
G S A W C S M H Z Y J C D U K Q W A U Q N V K T O S Y X
A V R J O H J Z O H Q M C R U N A G K X K F P U R J A E
W X O L V S U L F U R D N R X V B N Z T B L Q T H F J Z
```

# CHAPTER 8: CARBON OXIDES AND THE CARBON CYCLE

In chapter one, we started out with just the carbon atom. It was so easy. In the past few chapters we have looked at some very complicated molecules. Let's take a few steps backward now and look at some relatively simple molecules. We'll start with a carbon atom attached to two oxygens: *carbon dioxide*.

The empirical formula is $CO_2$.

ball and stick

$$O = C = O$$

structural formula

Yahoo! It's all downhill from here! No more hard chemistry!!

Ahh... refreshingly basic. No functional groups, no hydrophobic tails, no complicated geometry. The only tricky thing about carbon dioxide is that the bonds between the atoms must be double bonds, because carbon must have four bonds.

Carbon dioxide is a natural part of our world. Animals make it, and plants use it. Carbon dioxide and oxygen are traded back and forth from animals to plants.

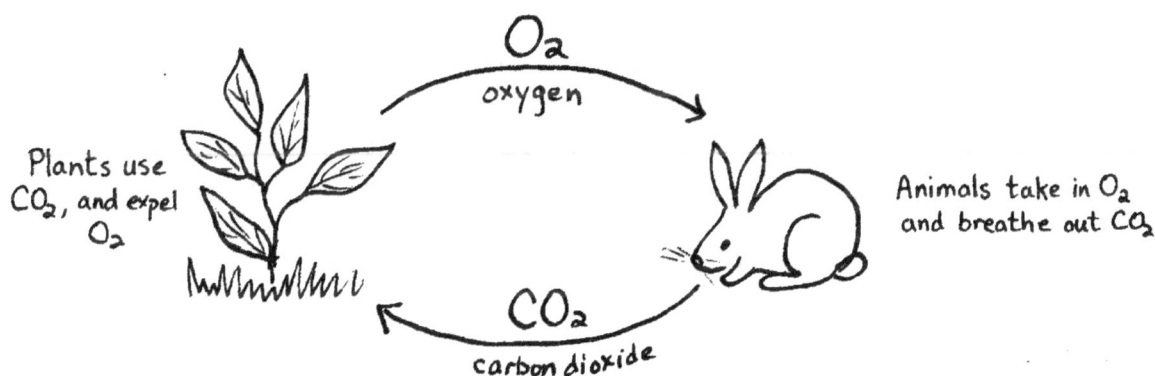

$O_2$
oxygen

Plants use $CO_2$, and expel $O_2$

Animals take in $O_2$ and breathe out $CO_2$

$CO_2$
carbon dioxide

Animals are not the only source of carbon dioxide, however. The earth itself produces large amounts of it. Volcanoes belch out huge amounts of carbon dioxide. Mt. Etna, in Italy, produces 35,000 tons per day. There are natural carbon dioxide vents at various points around the globe that put as much as 200 tons of $CO_2$ into the air every day. Researchers from Penn State University estimate that between the Italian cities of Naples and Florence, there are over 150 vents, each producing hundreds of tons of carbon dioxide per day.

Vehicles and factories also produce carbon dioxide, which is why studies are being done to try to determine if there is more $CO_2$ in the air than plants and ocean algae can absorb and use. Unfortunately, this has become not just a scientific question but also a political question. It's hard t to find "raw" data that has not already been interpreted for you. Discussion of this topic is beyond the scope of this book, so we are going to move on and look at the lighter side of carbon dioxide: carbonated beverages.

In order to mix with a liquid, a gas often has to undergo a chemical change. In this case, the $CO_2$ must combine with water molecules to make carbonic acid, $H_2CO_3$. After that happens, one of the hydrogen atoms will lose its proton, resulting in one loose proton and a "bicarbonate" ion. We might assume that a "bi-carbon-ate" ion will have 2 carbon atoms, since the Latin word root "bi" means "two." But no, the bicarbonate ion has only one carbon atom.

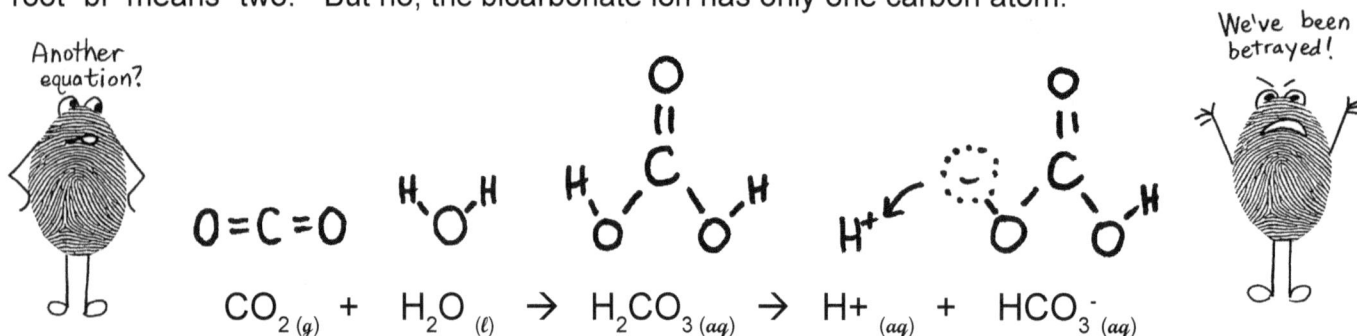

*Another equation?*

*We've been betrayed!*

$$O=C=O \quad + \quad H_2O \quad \rightarrow \quad H_2CO_3 \quad \rightarrow \quad H^+ \quad + \quad HCO_3^-$$

$$CO_{2\,(g)} \quad + \quad H_2O_{\,(\ell)} \quad \rightarrow \quad H_2CO_{3\,(aq)} \quad \rightarrow \quad H+_{\,(aq)} \quad + \quad HCO_3^-{}_{\,(aq)}$$

Those little script letters in parentheses aren't there to scare you by making the equation look more difficult. The letters are helpful hints that tell you whether the substance is a solid, liquid, or gas. The letters "aq" are short for "aqueous," which means that the substance isn't actually a liquid but is dissolved into a liquid.

If you look at a clear plastic bottle that contains a carbonated beverage, you won't see any bubbles. The equation above explains why. The liquid has dissolved ions of bicarbonate, not actual molecules of carbon dioxide. The beverage is kept under pressure (note how hard the bottle feels) so that the bicarbonate ions will stay dissolved. As soon as the bottle is opened and the liquid is no longer under pressure, the chemicals can reverse the equation and the bicarbonate can turn back into gaseous $CO_2$. Your body uses a similar process to transport $CO_2$ through your blood. When the dissolved $CO_2$ reaches the lungs, it turns into a gas again and you exhale it.

Chemists have a fancy name for this process of coming out of solution and turning back into a gas: "heterogeneous nucleation." The would-be gas molecules need a pre-existing microscopic gas pocket as a starting place were a gas bubble can form. This pre-existing pocket could be a tiny imperfection in the sides of the container. Even smooth glass will have tiny rough places you can't see, though it will have far fewer than a plastic bottle. This is why you see bubbles forming on the sides of a glass or bottle. Each bubble forms at a tiny nucleation site. You can disturb these sites and make the $CO_2$ go back into solution by tapping on the sides. (This trick works even with an unopened can that has been shaken. Flick your finter against the side of the can several times, and it won't explode when you open it.)

The secret to making a really bubbly beverage is to use very cold water (almost freezing temp) and very high pressure. Oddly enough, water can hold more $CO_2$ when it is cold than when it is hot. Normally, solutions can hold more "solute" (the stuff being dissolved) at higher temps. Not so with $CO_2$ in $H_2O$. While keeping the water very cold, you keep the air pressure in the space above the water very high. This will prevent the $CO_2$ from escaping and going back into the air.

*What are you doing?*

*Taking a step backward towards easier chemistry!*

Now let's take another step backward and look at the simplest of all carbon oxide molecules: carbon monoxide.

A carbon atom with only one oxygen attached to it is called *carbon monoxide*, *CO*.

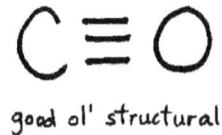

ye old ball 'n' stick model

good ol' structural

$$C \equiv O$$

How can this be? Carbon must have four bonds, and oxygen can only form two bonds. How can we have a molecule with three bonds? Chemists can't fully understand it or explain it. Sometimes it is drawn like this:

However it is drawn or explained, it's an anomaly. That means it's an exception to the rules. It shouldn't exist, but it does.

Carbon monoxide is often a byproduct of the combustion of fossil fuels. When octane (gasoline) is burned in engines, the chemical formula is something like this:

$$2 \, C_8H_{18} \; + \; 25 \, O_2 \; \rightarrow \; 18 \, H_2O \; + \; 16 \, CO_2$$
octane    oxygen        water    carbon dioxide

No carbon monoxide is formed in this reaction. The problem comes when there is not enough oxygen on the left side of the equation. What if there aren't 25 oxygens available for every two octanes? Some of those carbon dioxides end up with only one oxygen, making them into carbon monoxide.

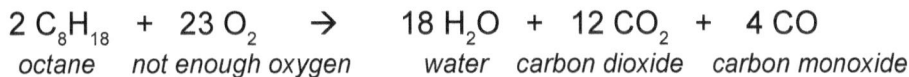

$$2 \, C_8H_{18} \; + \; 23 \, O_2 \; \rightarrow \; 18 \, H_2O \; + \; 12 \, CO_2 \; + \; 4 \, CO$$
octane  not enough oxygen    water  carbon dioxide  carbon monoxide

Carbon monoxide is bad stuff. Because carbon needs to have four bonds, it's out to fix the problem, no matter the cost. One thing carbon monoxide loves to bond with is the *hemoglobin* in your blood. It adores that iron atom sitting in the middle of your hemoglobin.

Oh, no! He's moving in on my territory!

Hellooo, iron! Wanna go for a ride with me?

Fe

Help! I need oxygen!

A cell calling to hemoglobin from off-stage.

The problem is that the iron atom has a job: transporting oxygen to your cells. But the CO doesn't care. It comes and attaches itself to the hemoglobin anyway. So off goes the hemoglobin to make its delivery to one of your cells. When the cell, which is desperately in need of oxygen, reaches out to take the oxygen off the hemoglobin, it finds… carbon monoxide?!! "I can't use this stuff!" the cell complains. "Go back to the lungs and get an oxygen for me!" The hemoglobin goes back to the lungs, but the carbon monoxide just won't get off. It's stuck. Crisis!

When this happens to a lot of your hemoglobin cells, it is called carbon monoxide poisoning, and it can kill you rather quickly. Since carbon monoxide has no smell, you can't rely on your nose to tell you if you are experiencing high levels of carbon monoxide. You don't know it's there until you start feeling sleepy. But feeling sleepy is not always unpleasant, so you don't immediately sense danger. Then you fall asleep and never wake up.

Since carbon monoxide can also be a byproduct of combustion in devices that heat homes, many homes are now equipped with carbon monoxide detectors. These devices work like smoke detectors, emitting a loud beeping noise.

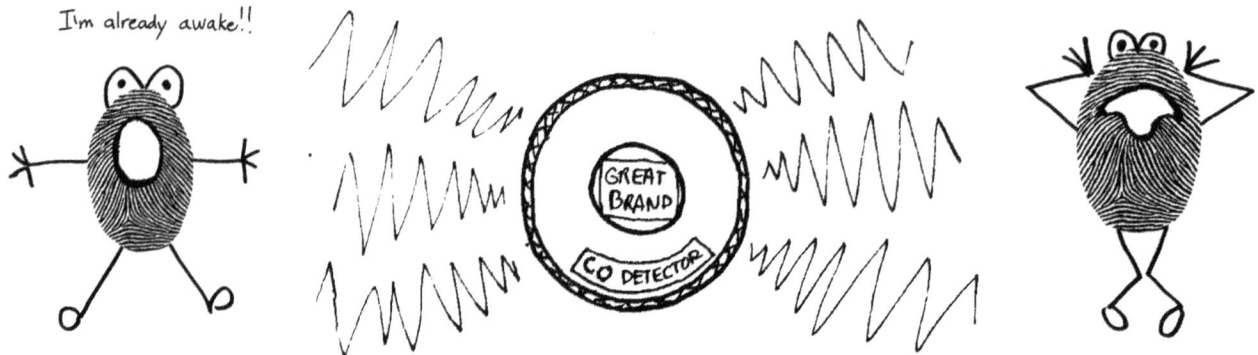

When carbon combines with three oxygens, we call it the **carbonate ion**, written like this: $CO_3^{2-}$ and drawn like this:

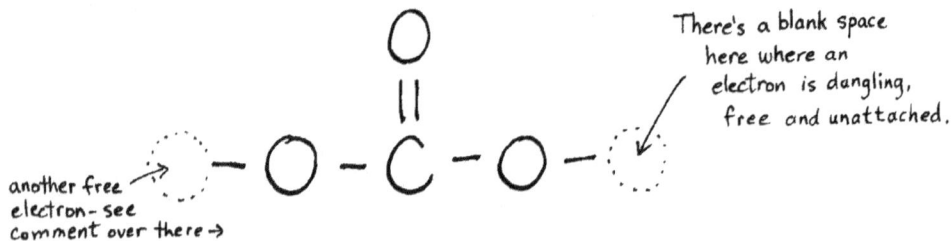

Carbon is happy, but two of the oxygens are not. Oxygen atoms need to make two bonds. These two oxygens have electrons dangling, unbonded. Since electrons have a negative electrical charge, these two unbonded electrons give this molecule an electrical charge of negative two. That's what the 2- means when we write $CO_3^{2-}$.

An atom or molecule with an electrical charge is called an *ion*. $CO_3^{2-}$ is called the *carbonate ion*, and is found in all sorts of places. The carbonate ion can "clean up" unwanted mineral ions from water.

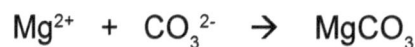

$$Mg^{2+} \; + \; CO_3^{2-} \; \rightarrow \; MgCO_3$$

Water with too many mineral ions, such as calcium or magnesium, is called "hard" water and can cause problems in water pipes. Notice the 2+ on magnesium. This means a positive electrical charge of 2. It's the exact opposite of $CO_3^{2-}$, which is why they are a perfect match for each other. The $MgCO_3$ will be more likely to flush out and not stick to the inside of pipes.

You will find the carbonate ions in the kitchen, too. Baking soda is the common name for the chemical compound called sodium bicarbonate: $NaHCO_3$. It's job in a recipe is to react with an mild acid, producing gas bubbles (carbon dioxide!) which make the food rise when baked, produced a light, fluffy texture.

In geology, the carbonate ion shows up in limestone, which is made of calcium carbonate, $CaCO_3$. Limestone covers vast areas of many continents, sometimes over a mile deep, and often found in alternating layers along with shale and sandstone. Areas with thick layers of limestone are likely to have caves or caverns. Scientists used to think that caves were formed solely by the action of carbonic acid seeping down through cracks in the rock. Carbonic acid is formed when carbon dioxide from the air is dissolved into rain water. Acids tend to dissolve things, so they assumed that carbonic acid had slowly and gradually dissolved the limestone to create the caves. New research, however, from caves in the western USA, suggests that a stronger acid, sulfuric acid, was involved in the formation of some limestone caves. The caves no longer contain sulfuric acid, just hints that it used to be there.

stalactite →

hibernating bats!

stalagmite →

The stalactites and stalagmites in these caves are reverse examples of the dissolving process. Water, saturated with dissolved calcium ions and carbonate ions, evaporates and leaves behind deposits of calcium carbonate.

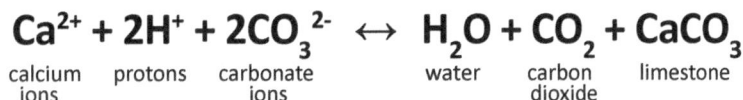

$$Ca^{2+} + 2H^+ + 2CO_3^{2-} \leftrightarrow H_2O + CO_2 + CaCO_3$$

| calcium ions | protons | carbonate ions | water | carbon dioxide | limestone |

Notice the arrow in the middle of this equation. It has a point on either end, which means the action can go in either direction. Note that when going from left to right, a molecule of $CO_2$ is formed for every molecule of limestone, $CaCO_3$. If you calculate all the $CO_2$ that would have been produced during the formation of the world's limestone, it would have made the atmosphere so toxic that no living thing could have survived. It may be necessary for scientists to rethink limestone formation and come up with a theory that doesn't put so much $CO_2$ into the atmosphere.

Ocean water contains a lot of dissolved carbon dioxide. There are also calcium ions in the water, so the ingredients needed to make calcium carbonate are very abundant. Many soft-bodied ocean animals are able to take calcium and carbonate out of the water and use it to build shells or exoskeletons. Mollusks (the snail family) make shells that are either two-sided (clams, mussels and oysters) or made of just one piece (snails, conches and whelks). Polyps (members of the jellyfish family) live in colonies that make elaborate solid shapes such as brain coral or staghorn coral. Chemically, there is not much difference between limestone and seashells. Because of the similarity, we might be tempted to think that limestone came from the shells of these ocean animals, but a quick review of the math involved, and a look at the crystals themselves, shows that most limestone rock must be inorganic (not from living things).

Most limestone may not have come from the shells of sea creatures, but another type of rock probably does: chalk. The famous white cliffs of Dover are made of a natural chalk, which is a very soft carbonate rock.

*I think this is my favorite page of the whole book!*

If you use a microscope to compare limestone with chalk, you will see a huge difference. Limestone crystals look inorganic. Chalk, on the other hand, looks like tiny bits of crushed shells left by microscopic single-celled sea creatures called coccolithophores *(cock-o-lith-o-fores)*. These extinct little animals were protozoans, distantly related to some tiny animals you may be familiar with: amoebas and paramecia. White cliffs of chalk are also found along the coasts of Denmark, on the other side of the Channel.

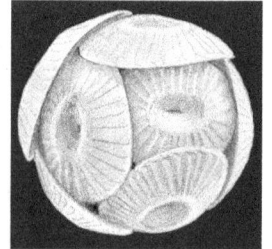

Calcium carbonate is an ingredient needed to make egg shells. Your teeth, also, are somewhat similar to a mineral related to limestone. This chemical similarity in such an array of seemingly unrelated things is a good introduction to our last topic: **the carbon cycle**.

A carbon atom is a carbon atom, no matter where it is found. A carbon atom found in limestone or in an eggshell is identical to a carbon atom found in polyethylene plastic, gasoline, or in food molecules such as sugars, proteins or fats. In fact, a carbon atom can switch from one type of molecule to another. When a plant takes in $CO_2$ from the air, it tears this molecule apart and uses the carbons and oxygens separately. It puts the carbon atom into a glucose (sugar) molecule. So the very same atom that was floating around in the atmosphere as carbon dioxide can end up in a molecule that an animal will eat.

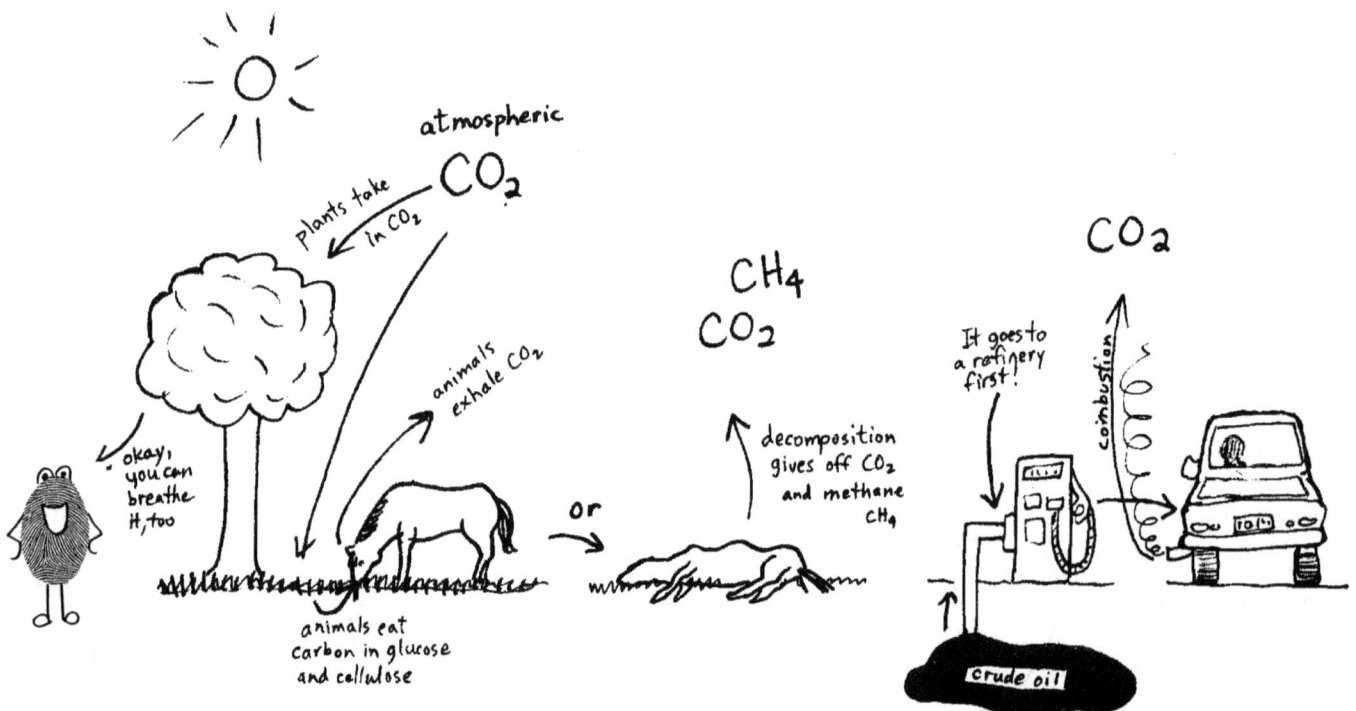

atmospheric $CO_2$

plants take in $CO_2$

animals exhale $CO_2$

okay, you can breathe it, too

animals eat carbon in glucose and cellulose

$CH_4$
$CO_2$

or

decomposition gives off $CO_2$ and methane $CH_4$

$CO_2$

It goes to a refinery first!

combustion

crude oil

64

The carbon atom will travel through the animal's body and might end up in other organic molecules, or be harvested for energy and then expelled through the lungs as carbon dioxide. If the carbon atom is incorporated into the body in a more permanent way (such as part of a DNA molecule), it might stay there until the animal dies. After death, bacteria will decompose the body and transfer the carbon atoms into either carbon dioxide or methane. If the carbon is released into the air as carbon dioxide, it is ready to start the whole cycle again, being taken in by a plant and converted to glucose. If the carbon ends up in methane, the gas can be collected and used as cooking fuel, or as a raw material to manufacture plastics.

Some carbon atoms have been stuck in one part of the cycle for a very long time. The carbons in crude oil were once part of glucose and starch molecules in living plants many years ago. The oil has been sitting, waiting to be discovered and used, and thus put back into the cycle. As we learned in chapter two, crude oil is turned into petroleum and gasoline for vehicles. The gasoline combusts in vehicles and is thus put back into the atmosphere as carbon dioxide.

$$2 C_8H_{18} + 25 O_2 \rightarrow 18 H_2O + 16 CO_2$$

We also learned that not all petroleum is used for fuel. Some of it goes to factories that make plastics, paints, or solvents. Some lucky carbon atoms will somehow get back into action, but other carbon atoms will end up in landfills where they will have to wait a long time to get back into the carbon cycle.

I don't think any of these things should be lying here.

Yeah, I think these are all recyclable!

---

REVIEW: Can you put the numbers 1 to 10 in the correct blanks?

1) _____ is great!
2) Nitrogen has _____ protons.
3) There are _____ carbon atoms in a butyric acid molecule.
4) There are _____ protons in a methyl group ($CH_3$).
5) There are _____ hydrogen atoms in butane.
6) "Meth" means _____.
7) Propane has _____ carbon atoms.
8) Naphthalene has _____ benzene rings.
9) A pyrrole ring is a ___-sided shape.
10) A benzene ring has _____ carbon atoms.

## Comprehension self-check

1) What kind of living organism needs $CO_2$? _____

2) What kind of living organism produces $CO_2$? _____

3) Name two non-living sources of $CO_2$: _____, _____

4) How many carbon atoms are in a bicarbonate molecule (despite the name)? _____

5) What is $H_2CO_3$? _____ (Helpful hint: Molecules that start with "$H_2$" are ALWAYS acids.)

6) What happens at a microscopic nucleation site?

   a) a molecule of carbonic acid forms     b) a bubble forms

   c) carbonic acid is turned into a bicarbonate ion and a free proton

7) Which will hold more dissolved $CO_2$, hot water or cold water? _____

8) Is there a way to reverse the situation once someone had shaken a bottle or can of carbonated beverage? _____

9) In addition to cold water, you also need _____ _____ to achieve maximum carbonation.

10) Carbon monoxide can be a by-product of:

   a) photosynthesis     b) carbonation     c) combustion     d) the carbon cycle

11) CO is dangerous in your body because it prevents _____ molecules from carrying oxygen to cells.

12) What is the first symptom of CO poisoning?

   a) vomiting     b) headache     c) pain in hands and feet     d) feeling sleepy

13) $CO_3^{2-}$ is called the _____ ion. (Remember, an ion is a "broken" molecule or atom that is no longer electrically balanced and therefore has a positive or negative charge.)

14) What is $NaHCO_3$? _____

   Where might you find this chemical? _____

15 What do you call water that has a lot of minerals such as calcium and magnesium?

   a) soft     b) hard     c) carbonated     d) acidic     e) mineralized

16) What is the chemical formula for limestone? _____

17) Which one of these does NOT contain calcium carbonate?

   a) limestone     b) egg shells     c) chalk     d) sea shells     e) carbonated beverages

18) Plants take carbon atoms out of $CO_2$ and use the process of photosynthesis to put the carbon atoms into molecules of _____.

19) Carbon atoms in petroleum might end as any of these EXCEPT:

   a) rubber     b) plastic     c) gasoline     d) paint     d) solvents     e) natural gas

20) What word do you use for a carbonated beverage?

   a) soda     b) pop     c) coke     d) soft drink     e) other _____

If you like trivia about words, try typing "map of words soda pop" into an Internet search engine. You will find maps showing where these words are commonly used. Most maps are of the USA but you will also be able to find at least one world map.

# CHAPTER 9: RADIOACTIVE CARBON

Let's go back and review the carbon atom again. Most carbon atoms have 6 protons and 6 neutrons in the nucleus, and 6 electrons in orbit around the nucleus. The number of protons determines what element an atom is, so if we add a proton to carbon's nucleus, giving it a total of 7 protons, it will instantly become a nitrogen atom. Adding neutrons, however, will not change the atom's identity. A carbon atom with 7 or 8 neutrons is still carbon.

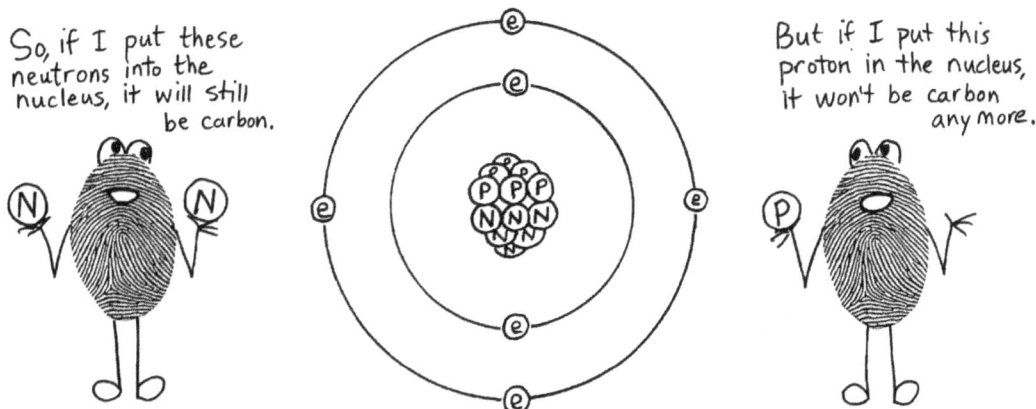

*So, if I put these neutrons into the nucleus, it will still be carbon.*

*But if I put this proton in the nucleus, it won't be carbon any more.*

The mass of an atom is the sum of the mass of all its parts. Mass is almost like weight, but scientists prefer to use the term mass because an object's mass will never change, whereas its weight might change depending on the location of the object. You probably know that if you weigh yourself on the moon, the scale will register only one sixth of your weight on earth. Your mass hasn't changed, but the gravity of the moon is one sixth of earth's gravity, so the moon pulls down on you less and therefore the scale registers less "weight." Weight is relative to where you are, but mass is not. As long as we are only talking about objects on the earth, we can use the words "mass" and "weight" interchangeably. But scientists like to be precise, so they prefer the term "mass."

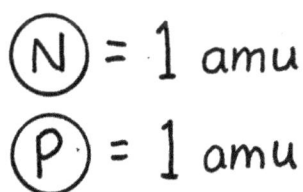

*I came all the way to the moon just to weigh myself!*

$$(N) = 1 \ amu$$

$$(P) = 1 \ amu$$

The units you use to weigh yourself are "pounds" or "kilograms." These units are far too large for measuring even a bacteria, let alone an atom. So scientists decided to come up with a special unit for weighing atoms: the **a**tomic **m**ass **u**nit, or **amu**. One amu is the mass of one proton or neutron; they both have the same mass so it doesn't matter which you choose. Electrons are so tiny that they add almost no mass to the atom. For all practical purposes we can ignore them and calculate mass just by adding the protons and neutrons.

99% of all carbon atoms have 6 protons and 6 neutrons, giving them a mass of 12 amu. Most of the remaining 1% will have 6 protons and 7 neutrons, giving a mass of 13 amu. A tiny fraction of that remaining 1% (one atom out of a trillion) will have 8 neutrons, giving it mass of 14. If we measure the mass of thousands of carbon atoms and then take the average of our results, we will get 12.01. You may have noticed that the mass numbers listed for most elements on the Periodic Table are decimal numbers. Most elements are like carbon, with individual atoms occasionally having more or less than the average number of neutrons.

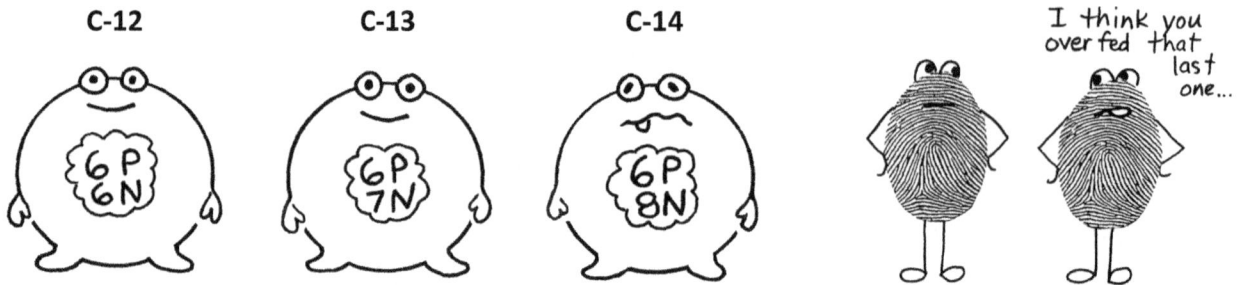

C-12   C-13   C-14

I think you over fed that last one...

Carbon-12, carbon-13 and carbon-14 are called *isotopes* of carbon. They are sometimes written like this: $^{12}C$, $^{13}C$, $^{14}C$. All elements have isotopes. Some isotopes are *stable* and, as the term suggests, won't fall apart. Some isotopes are *unstable* because their ratio of protons to neutrons is awkward. As atoms get larger, they require more and more neutrons compared to protons. Uranium, with 92 protons, must have at least 140 neutrons (and still isn't entirely stable). Smaller atoms, like carbon, have an ideal ratio of 50/50 (same number of protons and neutrons) but some of them can tolerate taking on an extra neutron. Carbon-12 and carbon-13 are very stable. Carbon-14 is unstable, but can act stable for many years. Some isotopes of very large atoms fall apart only a few seconds after receiving extra neutrons.

Our carbon-14 friend up there looks a little woozy. Neutron number 8 isn't sitting so well in his nucleus. Sooner or later, that 8th neutron will have to go. Atoms have their own version of vomiting; they can eject particles through a process called *decay*. The two most common types of decay are *alpha* and *beta* (Greek letters A and B). Any atom that will eventually decay is said to be *radioactive*. This term was invented around the turn of the 20th century when certain elements were discovered to give off **ra**ys of energy.

In *alpha decay*, the nucleus spits out an "alpha particle" made of 2 protons and 2 neutrons. In some cases, an atom undergoing alpha decay will also spit out some really dangerous energy called gamma rays. (Gamma rays are like high-energy x-rays.)

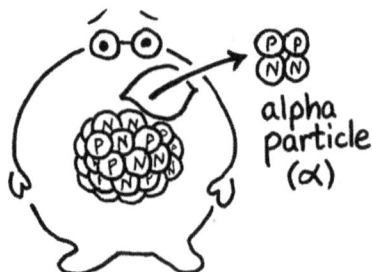

alpha particle (α)

In *beta decay,* a neutron turns into a proton, or a proton turns into a neutron. When a neutron turns into a proton, the atom spits out an electron. When a proton turns into a neutron, the atom spits out a positron (an electron with a positive charge instead of the usual negative charge).

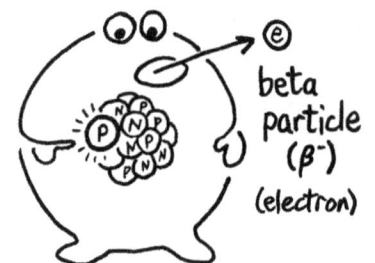

beta particle (β⁻) (electron)

Carbon-14 undergoes beta decay. The 8th neutron will turn into a proton, and the atom will give off an electron in the process. After the decay, the atom will have 7 protons and 7 neutrons, a nice 50/50 ratio. But wait... it changed its number of protons! All atoms with 7 protons are nitrogen atoms. Carbon has turned into nitrogen. However, this is not a surprise to anyone who knows how carbon-14 got its extra neutrons to begin with. Carbon-14 has merely gone back to its original identity.

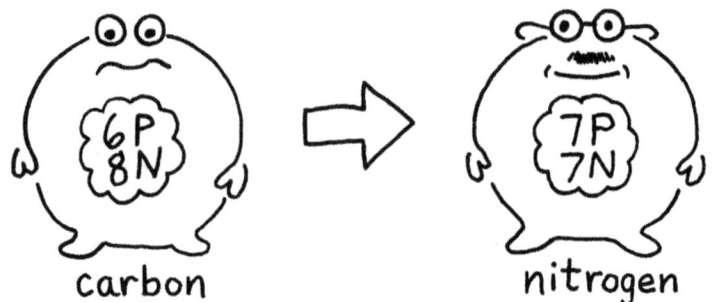

carbon          nitrogen

The primary (and perhaps only) source of the earth's naturally occurring carbon-14 is the upper atmosphere. High-energy cosmic rays from the sun strike the air molecules in the atmosphere. When these rays happen to strike the nucleus of an atom, they can dislodge a neutron and cause it to go flying off. If that loose neutron happens to collide with the nucleus of a nitrogen atom, it can dislodge and replace one of the protons. This causes the number of protons to go from 7 to 6. The nitrogen atom turns into a carbon atom, but the mass stays the same, at 14.

An atom of carbon-14 floating around in the atmosphere will often combine with two oxygen atoms to make a molecule of carbon dioxide. This "heavy" version of carbon dioxide (with its two extra neutrons) acts exactly the same as the regular $CO_2$. The heavy $CO_2$ will be taken up by plants and used in photosynthesis. The carbon-14 will then end up in a glucose molecule and might eventually be eaten by an animal or a person.

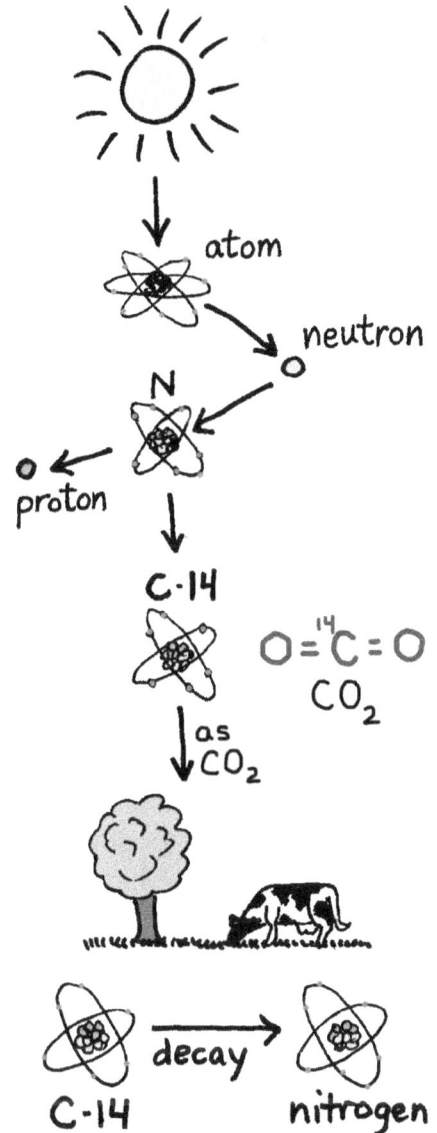

That means I'm about to eat a bunch of radioactive atoms!

I've read that bananas have radioactive potassium, too!

atom

neutron

N

proton

C-14

$O = \overset{14}{C} = O$
$CO_2$

as $CO_2$

C-14    decay    nitrogen

Your body contains millions of carbon-14 atoms in its various proteins, carbohydrates, fats, and DNA. Most of your carbon-14's will remain stable and not cause any problems. However, a small percentage will decay and turn into nitrogen. (About 1,000 of your C-14's will decay every second!) What would happen if one of the carbon atoms in a DNA molecule suddenly turned into a nitrogen atom? The DNA molecule contains important information for your cells. Changes in DNA are called mutations, and these changes are almost always harmful. A cell can only handle so many mutations before it becomes non-functional. The good news is that the beta particle (electron) that is released during beta decay isn't very powerful and is unlikely to cause much damage.

In past centuries, the atmosphere was the only source of carbon-14. Today, some of our carbon-14 has come from the burning of fossil fuels, from nuclear power plants, and from the testing of nuclear bombs. People who were alive during the 1950s, a decade when many atom bombs were tested, have higher amounts of C-14 in their bodies. If you carbon-dated these people, their radiocarbon age would be younger than their actual age.

In 1949, a few years after the discovery of C-14, a man named Willard Libby had an interesting idea. He thought that perhaps C-14 could be used to estimate the age of artifacts that had once been alive, such as bone, wood, leather, hair, and wool. Living plants and animals are constantly taking in carbon, some of which will be C-14. When a plant or animal dies, it stops taking in carbon. As the years roll by, the C-14 will gradually disappear as it decays into nitrogen. Libby reasoned that if we can measure the remaining amount of C-14 in a specimen, we can get a rough idea of how old it is.

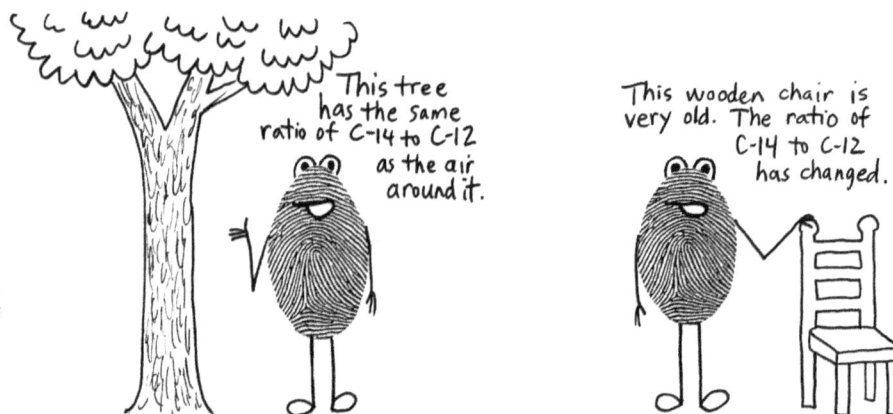

This tree has the same ratio of C-14 to C-12 as the air around it.

This wooden chair is very old. The ratio of C-14 to C-12 has changed.

Libby did experiments to determine the rate of decay for carbon-14. He could not predict when any individual atom would decay, but he was able to determine the *half-life* of carbon-14. The half-life is the length of time it takes for half of the radioactive atoms in an artifact to decay. The half-life of C-14 turned out to be about 5,730 years. For example, if a tree trunk had a million C-14 atoms in it when it was cut down, 5,730 years later it would have only half a million. In another 5,730 years, half of those would be gone, reducing the number to a quarter million. In another 5,730 years half of that would be gone, leaving one eighth of the original amount. After about 60,000 years, the amount of C-14 would be so small that it would not be measurable.

Libby made an assumption when he suggested this dating method. He assumed that the level of carbon-14 in the atmosphere has always been about what it is now. However, if the level has been a lot higher or a lot lower in the past, the age estimates for archaeological artifacts won't be accurate. Because we don't have C-14 data for the ancient past, this dating method is most useful with specimens that are relatively recent. For example, it is useful for dating wood or cloth artifacts from the Roman Empire or from the Middle Ages. The older the specimen, the less confident we can be about C-14 dating. However, this doesn't stop scientists from doing C-14 testing on very old things.

Recently, carbon-14 dating has stirred up controversy because C-14 has been found in all sorts of places where it "shouldn't be." Dinosaur bones, which are said to be millions of years old, have C-14 dates between 20,000 to 40,000 years old. (See page 82.) Also, C-14 has been detected in limestone, coal, oil, natural gas, fossils and diamonds, all of which are said to be millions of years old. Diamonds are the most surprising because there is no way they can possibly have been contaminated with any recent carbon. The carbon atoms in a diamond's lattice are locked in very tightly. No "fresh" carbon atoms can get in.

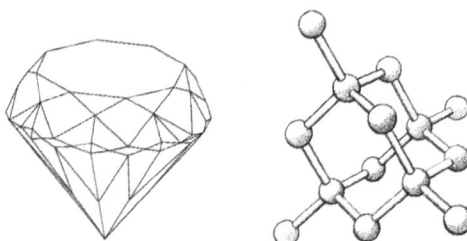

# HOW CAN A PROTON TURN INTO A NEUTRON?

In 1964, Murray Gell-Mann and George Zweig (working independently) both came up with similar theories to explain some of the odd behaviors of protons and neutrons. They both suggested that protons and neutrons were made of even smaller particles. The particles eventually became known as **quarks**, a term that Gell-Mann is said to have taken out of a strange novel by James Joyce that was popular in the early 1900s. A character in the book says, "Three quarks for Muster Mark!" Gell-Mann had already been thinking of naming the particles "quorks," so when he saw this phrase in the novel, he knew "quark" was the right choice, especially since the phrase mentioned the number three, and the new theory proposed that protons and neutrons were made of three particles.

In 1968, experiments were designed to test this new theory about quarks. Physicists fired electrons at protons in a machine called a particle accelerator. They watched how the electrons bounced off when they hit a proton. The results of the experiments seem to confirm this new theory. Protons and neutrons appeared to be made of "quarks."

There are six different types of quarks and they have very odd names: up, down, charm, strange, top and bottom. The quarks inside protons and neutrons are **up quarks** and **down quarks**. Not surprisingly, the symbols for these quarks are arrows pointing up or down. Protons are made of two up quarks and one down quark. Neutrons are made of two down quarks and one up quark.

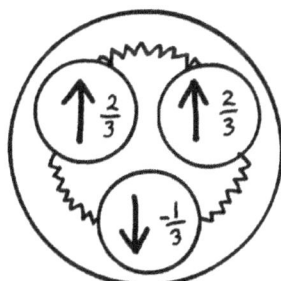

**PROTON**

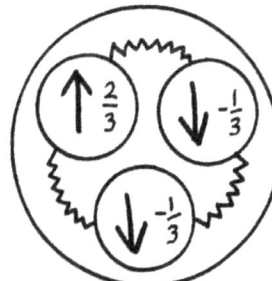

**NEUTRON**

Those zig-zaggy lines represent gluons, another type of subatomic particle. They hold the quarks together. Up quarks have an electrical charge of +2/3 and down quarks have a charge of -1/3. You can see how the fractions combine to give protons a charge of +1 and neutrons a charge of 0.

To change a proton into a neutron, all you need to do is replace an up quark with a down quark. How this happens is beyond the scope of this book. If you want to learn more, try starting with a TED-Ed video titled "What is the smallest thing in the universe?"

## Comprehension self-check

1) Does your mass change if you are standing on the moon? _____

2) What is the unit used to measure the mass (weight) of atoms? _____

3) What is the mass of a proton? _____ A neutron? _____

4) What percentage of all carbon atoms have 6 protons and 7 neutrons? _____

5) Only one in a _____ carbon atoms has 8 neutrons.

6) Atoms of the same element that differ in their number of neutrons are called
   a) allotropes    b) isotopes    c) isomers    d) ions

7) Any atom that will eventually undergo decay is said to be _____.

8) In alpha decay, an atom spits out an "alpha particle" that is made of two _____
   and two _____.

9) Carbon-14 undergoes beta decay and one of its _____ turns into a _____.

10) In beta decay, if a neutron turns into a proton, the atom ejects an _____.

11) Production of C-14 takes place in the upper atmosphere where cosmic rays from the sun hit atoms and dislodge _____, which then hit the nuclei of _____ atoms and turn them into carbon-14.

12) What happens to plants that take in "heavy" $CO_2$?
   a) Nothing    b) They become radioactive.    c) They die.    d) Their weight increases.

13) About how many C-14 atoms decay in your body every second? _____

14) Which of these did NOT generate more carbon-14?
   a) testing atomic bombs          b) burning fossil fuels (coal, oil, gas)
   c) running nuclear power plants    d) turning crude oil into plastics

15) What is the half-life of C-14? _____

16) After how many years will the C-14 level in any specimen be almost zero? _____

17) Can we be sure that the level of C-14 in the atmosphere has always been the same?_____

18) A proton is made of 2 _____ and 1 _____.

19) A neutron is made of 2 _____ and 1 _____.

20) How many kinds of quarks are there? _____

## Just for fun
Draw your own cartoon quarks. (up, down, strange, charm, top, bottom)

# REVIEW: Seven-letter words

1. ◯–E–◯–◯–◯–O–◯   9. ◯–E–◯–◯–◯–E
2. U–◯–◯–◯–◯–◯   10. ◯–◯–O–◯–◯–E
3. ◯–U–◯–◯–◯–◯   11. ◯–◯–◯–◯–O–◯–E
4. ◯–O–◯–◯–E–◯–◯   12. ◯–O–◯–◯–◯–E–◯
5. E–◯–◯–◯–◯–O–◯   13. ◯–◯–O–◯–◯–◯–E
6. ◯–◯–E–◯–O–◯–◯   14. ◯–A–◯–◯–◯–◯–◯
7. ◯–E–◯–◯–E–◯–◯   15. O–◯–◯–A–◯–◯–◯
8. ◯–◯–◯–O–◯–O–◯   16. ◯–◯–◯–O–◯–A–◯◯

Not all of the vowels have been given for every word. You may need to add a few more vowels.

CLUES (in no particular order):

This ring-shaped molecule has six carbon atoms.

This is the simplest hydrocarbon.

The variation of an atom that results when you change the number of neutrons.

This 5-sided ring is part of the indole molecule found in the Calabar bean.

Chemistry based on the element carbon is _____ chemistry.

A proton has two of this type of subatomic particle.

A complex acid (more than one functional group) used in food processing and baking.

The smallest ketone.

An alkane with three carbons.

The functional group (-OH)

This subatomic particle is made of two down quarks and one up quark.

A chain of many single units.

An element found in limestone along with the carbonate ion, $CO_3$.

An alcohol containing two carbon atoms and no double bonds.

Carbon _____ is needed by plants for photosynthesis.

A benzene ring with $CH_3$ attached. Can be extracted from the tolu tree.

# INDEX

# ANSWER KEY

## Chapter 1
1) protons, neutrons, electrons      2) electron cloud model, solar system model, ball and stick model
3) electron cloud     4) ball and stick     5) solar system     6) solar system     7) ball and stick
8) pea, football stadium, pinhead, upper decks      9) protons, atomic     10) "8 is great!"
11) It tries to borrow or share electrons with another atom, or atoms.     12) double bond
13) diamond, graphite, buckyball      14) dry lubricant    15) soccer ball (or "football" if you live outside USA)
16) graphene     17) without shape     18) coal, charcoal     19) plants      20) wood

## Chapter 2
1) methane      2) natural      3) yes      4) no      5) intestines      6) alkanes
7) names molecules     8) meth, eth, prop, but, pent, hex     9) gasoline (petrol) for cars     10) knocking
11) gases, liquids, solids      12) crude     13) refinery     14) distillation     15) cracking
16) chlorine, fluorine     17) chlorine, fluorine, carbon     18) mosquitoes, birds, reptiles
19) putting people to sleep before surgery     20) isomers
21) c     22) d     23) b     24) e     25) a

### "Hydrocarbon puzzle"
In order from top to bottom:
methane, refinery, distillation, fluorine, octane, cracking, alkane (or butane), propane, flammable, ozone, butane (or alkane)

### "Cross One Out"
1) rubbing alcohol     2) fermentation     3) ethyne     4) mountain tops
5) $C_2H_2$     6) helium     7) empirical formula     8) CFCs

## Chapter 3
1) single, double, triple     2) gases, solids
3) ripening fruit     4) saturated
5) unsaturated     6) 6
7) 40     8) single, double
9) aromatic     10) liver
11) toluene     12) 1
13) 1
14) plastic bottles, polyester fabric
15) 2     16) moth balls
17) moth     18) F
19) isomers
20) methane, natural gas

## Chapter 4
1) d     2) b
3) methanol, methyl alcohol, wood alcohol
4) one     5) ethanol, ethane
6) two     7) OH     8) COOH
9) vinegar     10) butter     11) ant
12) formalin (formaldehyde dissolved in water)
13) formaldehyde     14) CO
15) acetone     16) esters     17) banana
18) pineapple
19) putting people to sleep before surgery
20) gasoline (petrol)

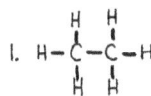

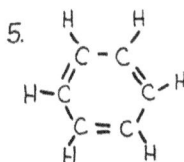

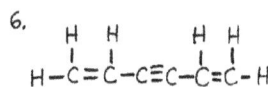

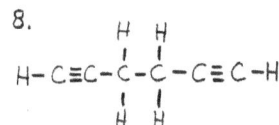

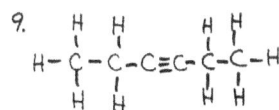

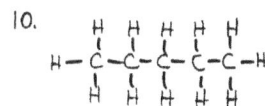

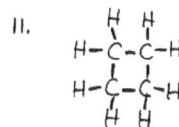

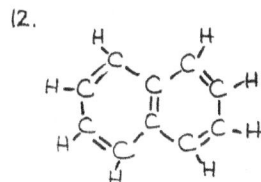

## Chapter 4, continued

**Crossword puzzle:**
Across:
2) methanol    4) isomer    5) aromatic    9) allotropes
10) octane    11) formic    12) saturated    13) three
15) distillation    16) empirical    19) chlorine    21) alkyne
22) neutrons    23) ethanol    24) graphite    25) graphene
Down:
1) carboxyl    3) xylene    4) IUPAC    6) amorphous
7) cracking    8) structural    14) hydrogen
16) ester    17) butyric    18) toluene
20) ethylene    21) alkene

**Matching (page 30):**
1) d    2) b    3) g    4) i    5) j
6) f    7) h    8) e    9) c    10) a

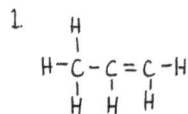

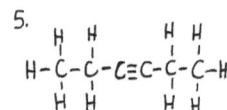

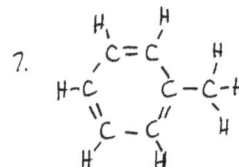

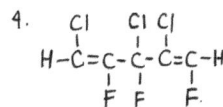

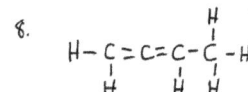

## Chapter 5
1) your choice: fumaric acid, tartaric acid, acetic acid          2) protons          3) benzoic acid
4) answers will vary, list on page 31          5) benzaldehyde          6) almond          7) ester
8) dynamite, relaxing blood vessels          9) animal fat, lye          10) wood ashes
11) b          12) no          13) hydrophobic          14) d          15) answers will vary, list on page 36
16) b          17) c          18) beans (Calabar in particular)
19) benzene, pyrrole          20) boiling temperatures
21) 4          22) alkene          23) isomers          24) allotropes          25) 8

**Review word puzzle on page 40:**
benzaldehyde, nitroglycerin, glaucoma, prostaglandin, empirical, pheromones, pyridine, benzoic, water, alcohol, jasmine, tolu, kiwi, tartaric, four, ether, hydrogens
"How long was the gypsy moth fooled by the pheromone trap? Just pheromo-ment!"

## Chapter 6
1) many, a single molecule (or molecular unit)          2) catalyst          3) 2
4) low density, high density          5) LDPE          6) LDPE
7) HDPE          8) LDPE          9) thermoplastic          10) 2
11) c          12) a          13) b          14) Styrofoam
15) polyvinylchloride          16) Teflon, non-stick          17) tusks, elephants
18) It exploded and/or caught on fire.          19) no          20) b

**REVIEW (pages 49, 50)**
1) B          2) F          3) D          4) E          5) A          6) C          7) H          8) G
9) G          10) C          11) A          12) D          13) E          14) H          15) B          16) F
17) fear of water, or water-hating          18) and 19) List any fact on page 36.
20) ester          21) yes          22) no smell, nothing          23) benzene ring          24) ant
25) preserve biological specimens
26) F          27) T          28) T          29) T          30) T
31) F          32) F          33) T          34) T          35) T          39-40)
36) F          3 7) T          38) T

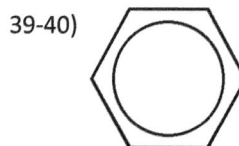

# Chapter 7

1) Central and South America    2) d    3) b    4) Charles Macintosh
5) Charles Goodyear    6) sulfur    7) d    8) Indonesia, Sri Lanka
9) Climate is too cold.    10) a    11) b    12) elastomers
13) c    14) 4    15-17) Answers will vary, see page 54.
18) Answers will vary depending on products you use.
water, veg. oil, cold syrup, shampoo, ketchup, petroleum jelly, Silly Putty
19) yes    20) no

## 1-800-4-REVIEW:

ELASTOMER    POLYMER    LATEX
RUBBER    CHICHLE    GUM
ISOPRENE    VISCOUS    SULFUR
BRAZIL    SAPODILLA    CROSSLINK    PRIESTLY
GOODYEAR    MACINTOSH    VULCANIZATION
INDONESIA

```
Y U J F U T G K W Y R I P C F V C B H D G U M T R X W C
L Q E V Y U Y G Z B P H I N D O N E S I A A M O R I S Y
Q J L D H D M W A L K U H Q E Y Z Q P O L Y M E R G W Y
L X A V N P Y N J U M O L I Y F C V H R I S O P R E N E
N K S Y W T J L K H W Y A G Z E D N V U E Z U Y D E E C
U F T M C K R K S O L L T S T S M F I K Q H G E U P P O
J W O Y M E E A J J K J E M R N B M S Y I U X V H H O R
R U M N T Y X T R L R W X M G F Z C C G H A R R U F E H
X F E K F O H K X Q V R L A I I S V O U H V U D Q B V J
G L R P R I E S T L Y E Z C W R L K U C T H B T Y H L N
V U L C A N I Z A T I O N I F H M P S O C R G Y Y E R P
F Z G A E J M C H I C L E N W Q B G C O R M K P K R F S
L Y O D R H H G E A V S X T Q P Q P K Y O Y B G E U I Q
H V O S A P O D I L L A E O A Y X H B W S I M A L B C W
E L D A R H V G I Q X S S S H X E N N G S A H N W B W D
F A Y V H T U S J R U T I H T Y E J Y U L R F P P E H R
V H E I K H E Y V E B R A Z I L D O P G I B S W M R H D
G S A W C S M H Z Y J C D U K Q W A U Q N V K T O S Y X
A V R J O H J Z O H Q M C R U N A G K X K F P U R J A E
W X O L V S U L F U R D N R X V B N Z T B L Q T H F J Z
```

# Chapter 8

REVIEW:
1) 8    2) 7    3) 4    4) 9    5) 10
6) 1    7) 3    8) 2    9) 5    10) 6

Comprehension self-check:
1) plants    2) animals
3) Answers will vary but can include: volcanoes,
$CO_2$ vents, factories, cars, combustion
4) 1    5) carbonic acid    6) b
7) cold    8) Tap on the side several times.
9) high pressure    10) c    11) hemoglobin
12) d    13) carbonate
14) sodium bicarbonate, baking soda or baking powder, baked goods such as cookies or muffins
15) b    16) $CaCO_3$    17) e    18) glucose (sugar)    19) a    20) Answers will vary.

# Chapter 9

1) no    2) atomic mass unit    3) 1 amu, 1 amu    4) about 1%
5) trillion    6) b    7) radioactive    8) protons, neutrons
9) neutrons, proton    10) electron    11) neutrons, nitrogen    12) a
13) 1,000    14) d    15) 5,730    16) 60,000
17) no    18) up quarks, down quark    19) down quark, up quarks    20) 6

## 7-Letter Word Review

1) neutron    2) upquark    3) fumaric    4) toluene    5) ethanol    6) acetone    7) benzene    8) alcohol
9) methane    10) isotope    11) pyrrole    12) polymer    13) dioxide    14) calcium    15) organic    16) propane

# Bibliography

*These are the main books I read to learn about organic chemistry.*

Chemistry for Changing Times, 8th edition, by John W. Hill and Doris K. Kolb, published by Prentice-Hall, Inc., Upper Saddle River, NJ. © 1998, ISBN 0-13-741786-1 (General chemistry textbook intended for high school or for college non-science majors.)

Chemistry of Carbon Compounds, 3rd edition, by David E. Newton, published by J. Weston Walch, Publisher, Box 658, Portland, Maine 04104.
© 1994, ISBN 0-8251-2487-5 (Organic chemistry text intend for advanced placement high school or entry level college.)

Contemporary Chemistry; A Practical Approach, by Leonard Saland, published by J. Weston Walch, Publisher, Box 658, Portland, Maine, 04104. ©1986, ISBN 0-8251-1799-2 (Leonard Saland is the chairman of the Physical Science Dept. at Louis Brandeis High School in New York.)

Penn State Earth and Mineral Sciences bulletin, Volume 66, 1997. (Article on research into Italy's carbon dioxide vents.)

"Hands-On Plastics," a scientific investigation curriculum put together by the American Plastics Council, 1801 K Street NW, Suite 701-L, Washington D. C., 20006. (Reference for experiment on identifying types of plastics.)

*I also used websites quite a bit. Here are the main ones I read, listed in order of how much I used them.*

www.Howitworks.com (one of my favorites!)
www.Chemed.chem.purdue.edu
www.Encyclopedia.com
www.Infoplease.com
www.chemistry.uakron.edu/genobc
www.elmhurst.ude/~chm/vchembook
www.medic8.com/healthguide/articles/exerciseanddiabetes
www.nexusresearchgroup.com/fun-science/fun.sci.htm
http://arbl/cvmbs/colostate.edu/hbooks/pathphys/digestion
www.post-gazette.com (for almond cookie recipe)
https://en.wikipedia.org/wiki/Martin_Kamen
https://en.wikipedia.org/wiki/Carbon-14
https://en.wikipedia.org/wiki/Willard_Libby
https://home.cern/news/news/physics/fifty-years-quarks

# SAMPLES OF LAB RESULTS FOR C-14 TESTING ON DINOSAUR BONES

These are just a few of the many tests that have been done. The ages they return are always between 10,000 to 40,000. The names of the labs and of the people who requested the tests have been obscured so that the author of this book will not be responsible for harrassment that might result. One of the labs has already decided to decline any further testing because of threats they have received from paleontologists.

---

a division of ▨▨▨ Enterprises, Inc

▨▨▨ • ▨▨▨ 1934 • USA
t ▨▨▨ f ▨▨▨ www.▨▨▨.om

### RADIOCARBON AGE DETERMINATION

**REPORT OF ANALYTICAL WORK**

| | |
|---|---|
| Our Sample No. | **GX-32678-AMS** |
| Your Reference: | see also GX-31950-AMS |
| Submitted by: | ▨▨▨ |

Date Received: 09/21/2006
Date Reported: 04/04/2007

Sample Name: **P-HI-(2)**

AGE = **22990 ± 130** $^{14}$C years BP ($^{13}$C corrected)

Description: Sample of charred bone

Pretreatment: The sample was cleaned of dirt and other foreign material and split into small pieces. It was then treated with hot dilute HCl to remove any carbonates; with 0.1N dilute NaOH to remove humic acids and other organic contaminants; and a second time with dilute HCl. After washing and drying, the sample was combusted to recover carbon dioxide for the analysis.

Comments:

$\delta^{13}C_{PDB}$ = **-18.4 ‰**

Notes: This date is based upon the Libby half life (5570 years) for $^{14}$C. The error is +/- 1 s as judged by the analytical data alone. Our modern standard is 95% of the activity of N.B.S. Oxalic Acid.

The age is referenced to the year A.D. 1950.

---

Center for Applied Isotope Studies

### RADIOCARBON ANALYSIS REPORT

May 31, 2011

▨▨▨
▨▨▨
▨▨▨

Dear Mr. ▨▨▨

Enclosed please find the results of carbon content analyses for the sample received by our laboratory on May 2, 2011.

| UGAMS # | Sample ID | Material | $^{14}$C age, years BP | $\delta^{13}$C, ‰ |
|---|---|---|---|---|
| 8824a | P-P-1 | bioapatite | 22020±50 | -5.4 |
| 8824carb | P-P-1 | carbonates | 4070±25 | -7.2 |

Bulk carbon content in the original bone sample-1.51%, N – 0.30%

The bone was cleaned and washed, using ultrasonic bath. After cleaning, the dried bone was gently crushed to small fragments. The crushed bone was treated with diluted 1N acetic acid to remove surface absorbed and secondary carbonates. Carbon dioxide from the secondary carbonates was collected and purified for analysis. The chemically cleaned sample was then reacted under vacuum with 1N HCl to dissolve the bone mineral and release carbon dioxide from bioapatite.

The resulting carbon dioxide was cryogenically purified from the other reaction products and catalytically converted to graphite using the method of Vogel et al. (1984) Nuclear Instruments and Methods in Physics Research B5, 289-293. Graphite $^{14}$C/$^{13}$C ratios were measured using the CAIS 0.5 MeV accelerator mass spectrometer. The sample ratios were compared to the ratio measured from the Oxalic Acid I (NBS SRM 4990). The sample $^{13}$C/$^{12}$C ratios were measured separately using a stable isotope ratio mass spectrometer and expressed as $\delta^{13}$C with respect to PDB, with an error of less than 0.1‰. The quoted uncalibrated dates have been given in radiocarbon years before 1950 (years BP), using the $^{14}$C half-life of 5568 years. The error is quoted as one standard deviation and reflects both statistical and experimental errors. The date has been corrected for isotope fractionation.

Sincerely,

---

a division of ▨▨▨ Enterprises, Inc.

▨▨▨ • ▨▨▨ :002 • USA
t ▨▨▨ f (6▨▨ www.▨▨▨.om

### RADIOCARBON AGE DETERMINATION

**REPORT OF ANALYTICAL WORK**

| | |
|---|---|
| Our Sample No. | **GX-32647** |
| Your Reference: | |
| Submitted by: | ▨▨▨ |

Date Received: 08/25/2006
Date Reported: 09/12/2006

Sample Name: **P-T-2**

AGE = **33830** $^{+2910}_{-1960}$ $^{14}$C years BP ($^{13}$C corrected)

Description: Sample of charred bone

Pretreatment: The charred bone fragments were cleaned of dirt and foreign material and treated with a benzene-acetone mixture to extract organic preservatives such as polyvinyl acetate (PVAc) and others. The sample was then dried at 130ºC. The dried sample was treated with hot 1N HCl to remove carbonates; with hot 0.1N NaOH to remove humic acids and other organic contaminants; and then again with dilute HCl to avoid absorbsion of CO₂ from the atmosphere. After washing and drying, the sample was combusted in flowing oxygen. The recovered CO₂ was converted to graphite and measured by accelerator mass spectrometry.

Comments: The sample was counted for an extended period of time.

$\delta^{13}C_{PDB}$ = **-16.6 ‰**

Notes: This date is based upon the Libby half life (5570 years) for $^{14}$C. The error is +/- 1 s as judged by the analytical data alone. Our modern standard is 95% of the activity of N.B.S. Oxalic Acid.

The age is referenced to the year A.D. 1950.

---

a division of ▨▨▨ Enterprises, Inc

▨▨▨ • ▨▨▨ 1934 • USA
t ▨▨▨ f ▨▨▨ www.g▨▨▨.com

### RADIOCARBON AGE DETERMINATION

**REPORT OF ANALYTICAL WORK**

| | |
|---|---|
| Our Sample No. | **GX-32678-AMS** |
| Your Reference: | see also GX-31950-AMS |
| Submitted by: | ▨▨▨ |

Date Received: 09/21/2006
Date Reported: 04/04/2007

Sample Name: **P-HI-(2)**

AGE = **22990 ± 130** $^{14}$C years BP ($^{13}$C corrected)

Description: Sample of charred bone

Pretreatment: The sample was cleaned of dirt and other foreign material and split into small pieces. It was then treated with hot dilute HCl to remove any carbonates; with 0.1N dilute NaOH to remove humic acids and other organic contaminants; and a second time with dilute HCl. After washing and drying, the sample was combusted to recover carbon dioxide for the analysis.

Comments:

$\delta^{13}C_{PDB}$ = **-18.4 ‰**

Notes: This date is based upon the Libby half life (5570 years) for $^{14}$C. The error is +/- 1 s as judged by the analytical data alone. Our modern standard is 95% of the activity of N.B.S. Oxalic Acid.

The age is referenced to the year A.D. 1950.

www.ingramcontent.com/pod-product-compliance
Lightning Source LLC
LaVergne TN
LVHW061336060426
835511LV00014B/1951